CALCULUS OF VARIATIONS

LIBRARY OF MATHEMATICS

edited by

WALTER LEDERMANN

D.Sc., Ph.D., F.R.S.Ed., Professor of
Mathematics, University of Sussex

CALCULUS
OF VARIATIONS

BY

A. M. ARTHURS

Department of Mathematics
University of York

ROUTLEDGE & KEGAN PAUL
LONDON AND BOSTON

First published in 1975
by Routledge & Kegan Paul Ltd
Broadway House, 68–74   arter Lane,
London EC4V 5EL and
9 Park Street,
Boston, Mass. 02108, USA
Printed in Hungary
© A. M. Arthurs 1975
ISBN 0 7100 7885 4
Library of Congress Catalog Card No. 74-77192

Contents

CONTENTS

Preface

The aim of this book is to provide an elementary introduction to variational problems. Some of these problems have a very long history, but the systematic study and development of the subject dates from the eighteenth century with the work of Euler and Lagrange.

The book opens with the elements of the Euler–Lagrange theory and continues with the extensions due to Hamilton and Jacobi and the connection with dynamic programming. Isoperimetric problems are also discussed and an introduction to direct methods is given in the final chapter.

Illustrative examples are included throughout the text, as well as applications to classical mechanics and boundary and eigen-value problems in differential equations.

I wish to thank Professor W. Ledermann and Dr W. A. Sutherland for reading the manuscript and for making valuable suggestions.

University of York   A. M. ARTHURS

CHAPTER ONE

Variational Problems

1. Introduction

Calculus of variations is a branch of analysis concerned with certain maximum or minimum problems. Its results can be used in many areas of mathematics and its applications, especially in the physical and engineering sciences. Some of the problems of calculus of variations have a long history going back to ancient Greek times, but the systematic study of variational problems dates from the eighteenth century with the work of Euler (1707–93) and Lagrange (1736–1813).

We begin by looking at a few typical examples of variational problems, as these will provide a background for the kind of questions we wish to discuss.

EXAMPLE 1.1. *The shortest path problem*

One of the earliest variational problems considered by the Greeks was to find the shortest distance between two points in a plane.

If the two points are $A(a, y_a)$ and $B(b, y_b)$, the length of the curve $Y = Y(x)$ joining them is given by the integral

$$J(Y) = \int_a^b \sqrt{(1+(Y')^2)}\, dx \quad (Y' = dY/dx). \quad (1.1)$$

We denote the integral (1.1) by $J(Y)$ to emphasise that it is a number J depending on the curve $Y = Y(x)$. In terms of (1.1),

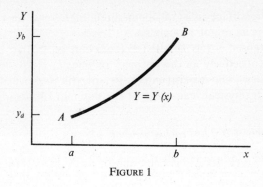

FIGURE 1

the shortest distance problem consists of finding the function or curve $Y = y$ which makes $J(Y)$ a minimum. Geometric intuition tells us that the curve y is the straight line joining A and B, though it is not entirely trivial to prove that this is the solution. More generally, if A and B are on some given surface, the shortest curve is a geodesic: for example, great circles on a sphere.

In ordinary calculus we meet the problem of finding *points* x at which a function $f(x)$ has maximum or minimum values. The above variational problem shows that in calculus of variations we try to find a *curve* or *function* y that minimises a quantity like $J(Y)$.

EXAMPLE 1.2. *The minimal surface problem*

If the curve $Y = Y(x)$ in Figure 1 is rotated about the x axis, a surface is generated which has area

$$J(Y) = 2\pi \int_a^b Y\sqrt{(1+(Y')^2)}\, \mathrm{d}x. \tag{1.2}$$

The minimal surface problem is to find the curve $Y = y$ which

2

makes the surface area (1.2) a minimum. In this case the required curve y is an arc of a catenary.

Generalised forms of this problem can be treated and are known collectively as the problem of Plateau (1801–83), after the physicist who experimented with soap films on wires, which provide a physical realisation of the minimal surface problem.

EXAMPLE 1.3. *The brachistochrone*

Our third example is a dynamical problem. It involves finding the shape of a smooth wire, joining two points in a vertical plane, down which a small bead will travel in minimum time. This is the brachistochrone problem of Galileo (1564–1642) and Bernoulli (1667–1748).

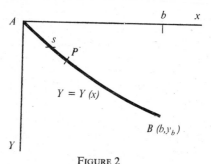

FIGURE 2

Using the diagram in Figure 2, we see that the speed of the bead at P is

$$v = \frac{\mathrm{d}s}{\mathrm{d}t}.$$

Hence the time of travel from A to B is

$$T(Y) = \int_{x=0}^{b} \frac{\mathrm{d}s}{v}. \qquad (1.3)$$

3

Now

$$ds = \sqrt{(1+(Y')^2)}\,dx, \qquad (1.4)$$

and from elementary dynamics, if the particle starts from rest at A,

$$v^2 = 2gY, \qquad (1.5)$$

where g is the acceleration due to gravity. So the integral to be minimised is

$$T(Y) = \int_0^b \frac{\sqrt{(1+(Y')^2)}}{\sqrt{(2gY)}}\,dx. \qquad (1.6)$$

The solution, that y is an arc of a cycloid, was obtained by several people, including Newton (1642–1727), Leibniz (1646–1716), James Bernoulli (1654–1705), and John Bernoulli (1667–1748).

EXAMPLE 1.4. *The isoperimetric problem*

Our fourth example is possibly the earliest variational problem considered by the Greeks. It is the geometrical problem of finding the closed plane curve of given length that encloses the largest area.

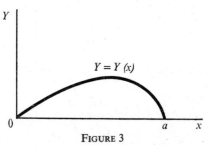

FIGURE 3

A simpler version of this, which was solved by Pappus (*c.* A.D. 300) concerns the area enclosed between the curve Y and an

4

interval of the x axis (Figure 3). In this case the problem is to find the function $Y = y$ which makes the area

$$J(Y) = \int_0^a Y \, dx \tag{1.7}$$

a maximum, subject to the length of the curve being fixed, that is

$$K(Y) = \int_0^a \sqrt{(1+(Y')^2)} \, dx = \text{constant}. \tag{1.8}$$

For this problem the intuitive answer, an arc of a circle, is correct. A condition like (1.8) is called an integral constraint, and the term *isoperimetric problem* is used to describe a variational problem involving an integral constraint.

Ancient legend has it that the land occupied by the city of Carthage was determined by the solution of an isoperimetric problem. The city was founded in 814 B.C. by the Phoenician princess Dido, who was given as much land as could be enclosed by the hide of an ox. With great ingenuity the princess cut up the hide into a very long leather thong and then formed it into a circle, thereby enclosing as much land as possible!

2. The fundamental problem

In this section we shall begin a systematic study of variational problems, and first we note that each of the examples 1.1–1.3 is a special case of the following more general problem. Suppose A and B are two fixed points with coordinates (a, y_a) and (b, y_b), and consider a set of curves

$$Y = Y(x) \tag{2.1}$$

joining A and B. Then we seek a member $Y = y$ of this set which minimises the integral

$$J(Y) = \int_a^b F(x, Y, Y') \, \mathrm{d}x. \tag{2.2}$$

This problem contains the above examples with

1. $F = \sqrt{(1 + (Y')^2)}$,

2. $F = 2\pi Y \sqrt{(1 + (Y')^2)}$,

3. $F = \dfrac{\sqrt{(1 + (Y')^2)}}{\sqrt{(2gY)}}$.

Now the curves $Y = Y(x)$ may be continuous or not, differentiable or not, and this affects the integral $J(Y)$. It will be simplest, however, in this introductory book to assume that the curves $Y = Y(x)$ are continuous and have continuous derivatives a suitable number of times. Once we specify the kind of curves we want, we have defined a set (or space) Ω of *admissible functions*. So the problem is to minimise the integral $J(Y)$ in (2.2) over the set of admissible functions which pass through the end points A and B. This problem is usually called the *fundamental problem* of the calculus of variations.

Extensions of this problem also arise, for example if the integrand F depends on Y'' as well as x, Y, Y', or if F contains two or more independent functions as in $F(x, Y_1, Y_2, Y_1', Y_2')$. These cases will be considered later.

We end this section by noting that the integral $J(Y)$ is an example of a *functional* which may be defined as follows:

DEFINITION. Let R be the real numbers and let Ω be a space of functions. Then the function $J : \Omega \to R$ is called a *functional*.

In terms of this definition we can say that calculus of variations is concerned with maxima and minima of functionals.

3. Maxima and minima

In elementary calculus we meet the idea of maxima and minima of functions[†] $f(x)$ of one variable x. A function $f(x)$ has a *minimum* at the interior point $x = a$ if

$$f(x) \geqslant f(a) \quad \text{for all } x \text{ near } a \tag{3.1}$$

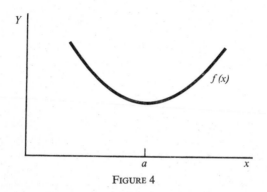

FIGURE 4

and a *maximum* is defined with the inequality reversed. To get some conditions from (3.1) we assume that $f(x)$ has a Taylor expansion about $x = a$ given by

$$f(a+h) = f(a)+hf'(a)+\tfrac{1}{2}h^2f''(a+\theta h), \quad 0 < \theta < 1. \tag{3.2}$$

Then (3.1) is equivalent to the statement

$$\Delta f \equiv f(a+h)-f(a) = hf'(a)+\tfrac{1}{2}h^2f''(a+\theta h) \geqslant 0 \tag{3.3}$$

for all h sufficiently close to zero. Now suppose $f'(a) \neq 0$. Then if h is small enough, the sign of Δf is determined by the

[†] See, for example, P. J. Hilton, *Differential Calculus*, Routledge & Kegan Paul (Library of Mathematics), London (1958).

7

sign of $hf'(a)$ and this can be made positive or negative by changing the sign of h. But $\Delta f \geqslant 0$, and so we must have $hf'(a) = 0$. Since h is arbitrary and non-zero, this implies that

$$f'(a) = 0. \tag{3.4}$$

This is the first necessary condition for an extremum (i.e. max. or min.). So we have

THEOREM 3.1. Let $f(x)$ be defined on the interval (x_0, x_1). Then if $f(x)$ has an extremum at $x = a$, $x_0 < a < x_1$, it follows that

$$f'(a) = 0.$$

The point $x = a$, where $f'(a) = 0$, is called a *critical* or *stationary point* of $f(x)$, and the value $f(a)$ is the *stationary value* of $f(x)$.

The stationary condition (3.4) does not tell us whether $f(x)$ has a maximum, or a minimum, or a point of inflection at $x = a$. To separate these out we go to the second derivative and from (3.3) we have

THEOREM 3.2. Suppose $f'(a) = 0$. Then

$$f''(a) > 0 \Rightarrow \Delta f \geqslant 0 \Rightarrow f(a) \text{ a minimum,} \tag{3.5}$$

and

$$f''(a) < 0 \Rightarrow \Delta f \leqslant 0 \Rightarrow f(a) \text{ a maximum.} \tag{3.6}$$

If $f''(a)$ changes sign for points near $x = a$, then this is a point of inflection.

These ideas of elementary calculus will now be taken over into calculus of variations, and for the moment we shall concentrate on the result corresponding to the stationary condition (3.4).

Consider the integral

$$J(Y) = \int_a^b F(x, Y, Y')\,dx \qquad (3.7)$$

of equation (2.2), and suppose that there is an admissible function $y(x)$ that minimises $J(Y)$. Then we have

$$J(Y) \geqslant J(y) \quad \text{for all } Y \in \Omega. \qquad (3.8)$$

This *minimum principle* is analogous to equation (3.1), and involves comparing J for various functions Y. If $J(Y)$ satisfies a maximum principle instead, we can turn it into a minimum principle by working with $-J(Y)$.

To compare $J(Y)$ with $J(y)$ we use the idea in (3.2), that is, we expand $J(Y)$ about $Y = y$ by taking Y in the form

$$Y(x) = y(x) + \varepsilon\xi(x), \qquad (3.9)$$

where ε is a small constant and $\xi(x)$ is an arbitrary function subject only to the condition that $y + \varepsilon\xi$ is admissible, that is, belongs to Ω. Equation (3.9) defines a set of varied curves about $y(x)$. Since all admissible functions Y, including y, go through A and B, it follows that

$$\xi(a) = 0, \; \xi(b) = 0, \qquad (3.10)$$

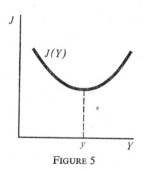

FIGURE 5

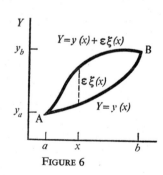

FIGURE 6

as shown in Figure 6. So the minimum principle (3.8) now becomes

$$J(y+\varepsilon\xi) \geqslant J(y) \quad \text{for all } y+\varepsilon\xi \in \Omega. \qquad (3.11)$$

To use (3.11) we adopt the procedure for (3.1), namely, we expand $J(y+\varepsilon\xi)$ in a Taylor series and have

$$J(y+\varepsilon\xi) = \int_a^b F(x, y+\varepsilon\xi, y'+\varepsilon\xi') \, dx$$

$$= \int_a^b \left\{ F(x, y, y') + \varepsilon\xi \frac{\partial F}{\partial y} + \varepsilon\xi' \frac{\partial F}{\partial y'} + 0(\varepsilon^2) \right\} dx$$

$$= J(y) + \delta J + 0(\varepsilon^2), \qquad (3.12)$$

where

$$\frac{\partial F}{\partial y} \equiv \left\{ \frac{\partial F}{\partial Y}(x, Y, Y') \right\}_{Y=y}$$

and

$$\delta J = \int_a^b \left\{ \varepsilon\xi \frac{\partial F}{\partial y} + \varepsilon\xi' \frac{\partial F}{\partial y'} \right\} dx$$

denotes the linear term in ε and is called the *first variation* of J. δJ is analogous to the linear term $hf'(a)$ in (3.2). If we integrate the ξ' term in δJ by parts and use (3.10) we find

$$\delta J = \int_a^b \varepsilon\xi \left\{ \frac{\partial F}{\partial y} - \frac{d}{dx} \frac{\partial F}{\partial y'} \right\} dx. \qquad (3.13)$$

Defining the brackets $\langle , \rangle$ by

$$\langle \phi, \psi \rangle = \int_a^b \phi(x)\,\psi(x) \, dx, \qquad (3.14)$$

and writing

$$J'(y) = \frac{\partial F}{\partial y} - \frac{d}{dx} \frac{\partial F}{\partial y'}, \qquad (3.15)$$

we see that

$$\delta J = \langle \varepsilon\xi, \ J'(y)\rangle. \tag{3.16}$$

In terms of this notation, the expansion (3.12) becomes

$$J(y+\varepsilon\xi) = J(y)+\langle \varepsilon\xi, \ J'(y)\rangle+0(\varepsilon^2), \tag{3.17}$$

which is analogous to the expansion (3.2) for a function $f(x)$

$$f(a+h) = f(a)+hf'(a)+0(h^2). \tag{3.18}$$

Equation (3.17) is a generalised Taylor expansion and in it $J'(y)$ is the *derivative* or *gradient* of $J(y)$, which for $J(Y)$ in (3.7) is given by equation (3.15).

We have seen in Theorem 3.1 that $hf'(a) = 0$ is a necessary condition for an extremum of $f(x)$, and by the same argument we have:

THEOREM 3.3. A necessary condition for $J(Y)$ to have an extremum at $Y = y$ is that

$$\delta J = \langle \varepsilon\xi, \ J'(y)\rangle = 0 \tag{3.19}$$

for all admissible $\varepsilon\xi$.

The curve $Y = y$ where $\delta J = 0$ is called a *critical curve*, or *extremal*, and the corresponding value of $J(y)$ is called a *stationary value* of the integral.

As in the case of ordinary functions, the stationary condition (3.19) does *not* tell us whether $J(Y)$ has a maximum, or a minimum, or a point of inflection for $Y = y$. To determine these properties, higher order terms must be considered, and for variational problems in general these are quite complicated. We shall look at some simple cases later, in Chapter 3. For the moment we shall concentrate on the stationary condition (3.19).

4. Euler–Lagrange equation

We now turn to the consequences of the stationary condition

$$\delta J = 0,$$

that is,

$$\langle \varepsilon\xi, J'(y)\rangle = 0 \tag{4.1}$$

for all admissible $\varepsilon\xi$. By analogy with the result for ordinary functions (Theorem 3.1), we would expect (4.1) to imply the condition

$$J'(y) = 0. \tag{4.2}$$

This is indeed correct and follows from the

Euler–Lagrange lemma. If $u(x)$ is continuous in $[a, b]$, and if

$$\langle \xi, u\rangle \equiv \int_a^b \xi(x)\, u(x)\, \mathrm{d}x = 0 \tag{4.3}$$

for every continuous function $\xi(x)$ such that $\xi(a) = 0$, $\xi(b) = 0$, then

$$u(x) = 0 \quad \text{for all } x \text{ in } [a, b]. \tag{4.4}$$

Proof. We prove this by assuming that $u(x) \neq 0$, say $u(x)$ positive, at some point in $[a, b]$. Then by continuity, $u(x)$ is positive in some interval $[x_0, x_1]$ containing this point and contained in $[a, b]$. If we set

$$\xi(x) = \begin{cases} (x-x_0)^2\,(x_1-x)^2, & x_0 < x < x_1, \\ 0 & \text{otherwise,} \end{cases} \tag{4.5}$$

then this $\xi(x)$ satisfies the conditions of the lemma. However,

$$\int_a^b \xi(x)\, u(x)\, \mathrm{d}x = \int_{x_0}^{x_1} (x-x_0)^2\,(x_1-x)^2\, u(x)\, \mathrm{d}x > 0$$

12

since the integrand is positive in (x_0, x_1). This contradicts (4.3), and so $u(x)$ is not non-zero, which proves the result.

Since $\xi(x)$ in (4.1) satisfies the conditions of the lemma, and since $J'(y)$ is a function of x like $u(x)$, it follows from the lemma that

$$J'(y) = 0. \tag{4.6}$$

In terms of the definition in equation (3.15), this means that

$$\frac{\partial F}{\partial y} - \frac{\mathrm{d}}{\mathrm{d}x} \frac{\partial F}{\partial y'} = 0, \tag{4.7}$$

which is known as the *Euler–Lagrange equation*. We have therefore proved

THEOREM 4.1. A necessary condition for the functional

$$J(Y) = \int_a^b F(x, Y, Y') \, \mathrm{d}x, \quad Y(a) = y_a, Y(b) = y_b,$$

to have an extremum for the function y is that y be a solution of

$$J'(y) = 0,$$

that is

$$\frac{\partial F}{\partial y} - \frac{\mathrm{d}}{\mathrm{d}x} \frac{\partial F}{\partial y'} = 0 \quad a \leqslant x \leqslant b,$$

with

$$y(a) = y_a, \quad y(b) = y_b.$$

This is the basic Euler–Lagrange variational principle and is the condition for the integral $J(Y)$ to be stationary at the critical curve $Y = y$.

We now apply this theorem to find the critical curves y for some of the examples mentioned in section 1.

EXAMPLE 4.1. *Shortest path problem*

For this problem

$$F(x, Y, Y') = \sqrt{(1+(Y')^2)}, \qquad (4.8)$$

and so the Euler–Lagrange equation (4.7) for the critical curve y is

$$0 = \frac{\partial F}{\partial y} - \frac{d}{dx}\frac{\partial F}{\partial y'} = -\frac{d}{dx}\left\{\frac{y'}{\sqrt{(1+y'^2)}}\right\}, \qquad (4.9)$$

which gives

$$y' = \text{constant},$$

or

$$y = \alpha x + \beta. \qquad (4.10)$$

Here α and β are constants determined from the boundary conditions $y(a) = y_a$ and $y(b) = y_b$. The result (4.10) shows that the critical curve is a straight line. This is the curve which makes $J(Y)$ stationary, and further investigation is needed to prove that it actually produces a minimum, though this would be expected from the geometry of the problem.

EXAMPLE 4.2. *Minimal surface problem*

For this problem

$$F(x, Y, Y') = 2\pi Y \sqrt{(1+(Y')^2)}, \qquad (4.11)$$

and the corresponding Euler–Lagrange equation for the critical curve y is

$$\sqrt{(1+y'^2)} - \frac{d}{dx}\left\{\frac{yy'}{\sqrt{(1+y'^2)}}\right\} = 0. \qquad (4.12)$$

Equation (4.12) leads to

$$(1+y'^2)^{-3/2}\{1+y'^2-yy''\} = 0 \qquad (4.13)$$

14

and so for finite y' in $a < x < b$ we obtain

$$1+y'^2 = yy''. \tag{4.14}$$

Using the identity $y'' = y'\mathrm{d}y'/\mathrm{d}y$, we may write (4.14) in separable form

$$\frac{\mathrm{d}y}{y} = \frac{1}{2}\frac{\mathrm{d}(y'^2)}{1+y'^2}.$$

Integrating once yields

$$y = C(1+y'^2)^{1/2}, \ C = \text{const.} \tag{4.15}$$

On solving for y' and integrating again we find that

$$y = C \cosh\left(\frac{x+C_1}{C}\right). \tag{4.16}$$

This critical curve is an arc of a *catenary*, where the constants C and C_1 are determined by the boundary conditions $y(a) = y_a$ and $y(b) = y_b$. The surface of revolution generated by this curve is called a *catenoid*. As in example 4.1, further investigation is needed to show that the critical curve (4.16) produces a minimum for $J(Y)$.

To end this section on the Euler–Lagrange equation we consider a generalisation of the fundamental problem in which the basic integral J depends on more than one function. Suppose $Y_1, Y_2, \ldots, Y_n$ are n independent functions of x and consider the problem of finding the set of curves $Y_1 = y_1, Y_2 = y_2, \ldots, Y_n = y_n$ which minimises the integral

$$J(Y_1, Y_2, \ldots, Y_n) = \int_a^b F(x, Y_1, \ldots, Y_n, Y_1' \ldots, Y_n') \, \mathrm{d}x. \tag{4.17}$$

Here each curve Y_j goes through fixed points A_j and B_j given by

$$Y_j(a) = y_{ja}, \ Y_j(b) = y_{jb}, \quad j = 1, \ldots, n. \tag{4.18}$$

15

Taking

$$Y_j = y_j(x) + \varepsilon \xi_j(x), \qquad (4.19)$$

where ξ_j vanishes at the end points, and using the Taylor expansion, we find that the first variation of J is

$$\delta J = \sum_{j=1}^{n} \langle \varepsilon \xi_j, J_j'(y_1, \ldots, y_n) \rangle, \qquad (4.20)$$

with

$$J_j'(y_1, \ldots, y_n) = \frac{\partial F}{\partial y_j} - \frac{\mathrm{d}}{\mathrm{d}x} \frac{\partial F}{\partial y_j'} \quad (j = 1, \ldots, n). \quad (4.21)$$

For an extremum we must satisfy the stationary condition $\delta J = 0$ (Theorem 3.3) and by (4.20) this means

$$\sum_{j=1}^{n} \langle \varepsilon \xi_j, J_j' \rangle = 0. \qquad (4.22)$$

Now the $\varepsilon \xi_j$ are all independent and hence (4.22) implies that

$$\langle \varepsilon \xi_j, J_j' \rangle = 0, \quad (j = 1, \ldots, n). \qquad (4.23)$$

Finally, using the Euler–Lagrange lemma in (4.3) and (4.4), we obtain the condition

$$J_j' = 0, \quad (j = 1, \ldots, n), \qquad (4.24)$$

that is

$$\frac{\partial F}{\partial y_j} - \frac{\mathrm{d}}{\mathrm{d}x} \frac{\partial F}{\partial y_j'} = 0, \quad (j = 1, \ldots, n). \qquad (4.25)$$

Thus the critical curves are found by solving the *system of n Euler–Lagrange equations* (4.25) subject to the boundary conditions (4.18).

5. Exercises

1. Find the critical curves of the functionals

(i) $\int_0^1 y'\,dx$, (ii) $\int_0^1 yy'\,dx$, (iii) $\int_0^1 xyy'\,dx$,

where in each case $y(0) = 0$, $y(1) = 1$.

2. Find the critical curves of the functionals

(i) $\int_a^b (y^2 + y'^2 - 2y \sin x)\,dx$, (ii) $\int_a^b (y^2 - y'^2 - 2y \sin x)\,dx$,

(iii) $\int_a^b (y^2 - y'^2 - 2y \cosh x)\,dx$, (iv) $\int_a^b (y^2 + y'^2 + 2ye^x)\,dx$.

3. Find the general solution of the Euler–Lagrange equation for the functional

$$\int_a^b f(x)\,(1 + y'^2)^{1/2}\,dx,$$

and investigate the special cases (i) $f(x) = x^{1/2}$, and (ii) $f(x) = x$.

4. Find the critical curves for

(i) $\int_{-1}^1 \{y(1 + y'^2)\}^{1/2}\,dx$, $y(-1) = y(1) = b > 0$.

(ii) $\int_a^b \dfrac{1 + y^2}{y'^2}\,dx$, $y(a) = A$, $y(b) = B$.

5. Show that equation (4.16) follows from equation (4.15).

6. Find the critical curve for the brachistochrone problem described in example 1.3.

7. Establish the result in equation (4.20).

17

8. Show that the Euler–Lagrange equation for the functional

$$J(Y) = \int_a^b F(x, Y, Y', Y'') \, dx$$

is

$$\frac{\partial F}{\partial y} - \frac{d}{dx} \frac{\partial F}{\partial y'} + \frac{d^2}{dx^2} \frac{\partial F}{\partial y''} = 0.$$

CHAPTER TWO

Some Extensions

---

6. Canonical Euler equations

At the end of section 4 we derived the Euler–Lagrange equations

$$\frac{\partial F}{\partial y_j} - \frac{\mathrm{d}}{\mathrm{d}x} \frac{\partial F}{\partial y_j'} = 0, \quad j = 1, \ldots, n, \tag{6.1}$$

which give the critical curves $y_1, \ldots, y_n$ for the functional

$$J(Y_1, \ldots, Y_n) = \int_a^b F(x, Y_1, \ldots, Y_n, Y_1', \ldots, Y_n') \, \mathrm{d}x. \tag{6.2}$$

The equations (6.1) form a system of n second-order differential equations, and we now propose to rewrite this as a system of $2n$ first-order differential equations. First we introduce a variable

$$p_i = \frac{\partial F}{\partial y_i'}, \quad i = 1, \ldots, n. \tag{6.3}$$

This p_i is said to be the variable *conjugate* to y_i. We suppose that (6.3) can be solved to give y_i' as a function of x, y_j and p_j ($j = 1, \ldots, n$). Then it is possible to define a new function $H(x, y_1, \ldots, y_n, p_1, \ldots, p_n)$ by the equation

$$H(x, y_1, \ldots, y_n, p_1, \ldots, p_n)$$

$$= \sum_{i=1}^{n} p_i y_i' - F(x, y_1, \ldots, y_n, y_1', \ldots, y_n'). \tag{6.4}$$

19

The function H is called the *Hamiltonian* corresponding to equation (6.2). We now consider the differential of H which, from (6.4), is given by

$$dH = \sum_{i=1}^{n} (p_i\, dy_i' + y_i'\, dp_i) - \frac{\partial F}{\partial x}\, dx$$

$$- \sum_{i=1}^{n} \left(\frac{\partial F}{\partial y_i}\, dy_i + \frac{\partial F}{\partial y_i'}\, dy_i' \right)$$

$$= -\frac{\partial F}{\partial x}\, dx + \sum_{i=1}^{n} \left(y_i'\, dp_i - \frac{\partial F}{\partial y_i}\, dy_i \right), \qquad (6.5)$$

the terms in dy_i' cancelling because of (6.3). It follows from this that

$$y_i' = \frac{\partial H}{\partial p_i}, \qquad -\frac{\partial F}{\partial y_i} = \frac{\partial H}{\partial y_i},$$

or

$$\frac{dy_i}{dx} = \frac{\partial H}{\partial p_i}, \qquad -\frac{dp_i}{dx} = \frac{\partial H}{\partial y_i}, \qquad i = 1, \ldots, n, \qquad (6.6)$$

where we have used (6.1) and (6.3). Equations (6.6) are the *canonical Euler equations* associated with the integral (6.2).

These canonical equations can be obtained in another interesting way. If we regard the functions Y_i and P_i as *independent* variables, we can now define a new functional, analogous to (6.2), by

$$I(Y_1, \ldots, Y_n, P_1, \ldots, P_n)$$

$$= \int_a^b \left\{ \sum_{j=1}^{n} P_j Y_j' - H(x, Y_1, \ldots, Y_n, P_1, \ldots, P_n) \right\} dx. \qquad (6.7)$$

Then, by (6.1), the corresponding Euler–Lagrange equations are

$$\left(\frac{\partial}{\partial y_i} - \frac{d}{dx}\frac{\partial}{\partial y_i'} \right) \left\{ \sum_{j=1}^{n} p_j y_j' - H \right\} = 0, \qquad i = 1, \ldots, n, \qquad (6.8)$$

20

and

$$\left(\frac{\partial}{\partial p_i} - \frac{\mathrm{d}}{\mathrm{d}x}\frac{\partial}{\partial p_i'}\right)\left\{\sum_{j=1}^{n} p_j y_j' - H\right\} = 0, \quad i = 1, \ldots, n. \quad (6.9)$$

Working out these expressions, we find that

$$-\frac{\partial H}{\partial y_i} - \frac{\mathrm{d}p_i}{\mathrm{d}x} = 0,$$

and

$$y_i' - \frac{\partial H}{\partial p_i} = 0,$$

or

$$\frac{\mathrm{d}y_i}{\mathrm{d}x} = \frac{\partial H}{\partial p_i}, \quad -\frac{\mathrm{d}p_i}{\mathrm{d}x} = \frac{\partial H}{\partial y_i}, \quad i = 1, \ldots, n. \quad (6.10)$$

These are just the canonical equations given in (6.6). They form a system of $2n$ first-order differential equations, and their solution $y_1, \ldots, y_n, p_1, \ldots, p_n$ gives the critical curves for the functional I in equation (6.7).

The canonical system of equations in (6.10) will be used later in the book.

EXAMPLE 6.1. To illustrate these canonical equations we take $n = 1$ and consider

$$J(Y) = \int_a^b (\alpha Y'^2 + \beta Y^2)\, \mathrm{d}x. \quad (6.11)$$

For this

$$F(x, y, y') = \alpha y'^2 + \beta y^2$$

and

$$p = \frac{\partial F}{\partial y'} = 2\alpha y' \Rightarrow y' = p/2\alpha.$$

21

The Hamiltonian H is, by (6.4),

$$H = py' - F = \frac{p^2}{4\alpha} - \beta y^2.$$

From (6.6) we find that the canonical equations are

$$\frac{dy}{dx} = \frac{\partial H}{\partial p} = \frac{p}{2\alpha}; \quad -\frac{dp}{dx} = \frac{\partial H}{\partial y} = -2\beta y. \quad (6.12)$$

The ordinary Euler–Lagrange equation for (6.11) is

$$\frac{\partial F}{\partial y} - \frac{d}{dx}\frac{\partial F}{\partial y'} = 0 \Rightarrow 2\beta y - \frac{d}{dx}(2\alpha y') = 0,$$

which is equivalent to (6.12).

7. Variable end points

So far we have considered variational problems in which the admissible functions Y take prescribed values at the end points $x = a$ and $x = b$. More general cases can arise, however, when values at the end points are not imposed on the functions, and we now look at these.

CASE 7.1. *End points variable in y direction*

Consider the problem of finding the function $Y = y$ which minimises the integral

$$J(Y) = \int_a^b F(x, Y, Y')\, dx, \quad (7.1)$$

where the functions Y are defined on $a \leqslant x \leqslant b$ but do not have values prescribed at the end points. By Theorem 3.3, a

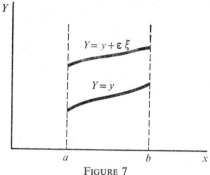

FIGURE 7

necessary condition for $J(Y)$ to have a minimum at $Y = y$ is that

$$\delta J\,(y,\ \varepsilon\xi) = 0 \qquad (7.2)$$

for all admissible $\varepsilon\xi$. To use this we require the first variation δJ, which is found from (7.1) by taking

$$J(y + \varepsilon\xi) = \int_a^b F(x, y + \varepsilon\xi, y' + \varepsilon\xi')\,\mathrm{d}x$$

$$= \int_a^b \left\{ F(x, y, y') + \varepsilon\xi\frac{\partial F}{\partial y} + \varepsilon\xi'\frac{\partial F}{\partial y'} + 0(\varepsilon^2) \right\}\,\mathrm{d}x.$$

Hence

$$\delta J = \int_a^b \left\{ \varepsilon\xi\frac{\partial F}{\partial y} + \varepsilon\xi'\frac{\partial F}{\partial y'} \right\}\,\mathrm{d}x, \qquad (7.3)$$

as seen earlier in section 3. Integrating the term in ξ' by parts, we obtain

$$\delta J = \int_a^b \varepsilon\xi \left\{ \frac{\partial F}{\partial y} - \frac{\mathrm{d}}{\mathrm{d}x}\frac{\partial F}{\partial y'} \right\}\,\mathrm{d}x + \left[\varepsilon\xi\frac{\partial F}{\partial y'} \right]_a^b. \qquad (7.4)$$

For fixed end points, ξ is zero at $x = a$ and $x = b$, and (7.4) reduces to the expression in equation (3.13). However, in the

23

present case the end points need not be fixed (Figure 7), and this means that ξ is *arbitrary* at $x = a$ and $x = b$.

Now the stationary condition $\delta J = 0$ must hold for all admissible functions Y, and we can separate these functions into four sets, namely functions such that

(i) $\xi(a) = 0$, $\xi(b) = 0$,   (iii) $\xi(a) \neq 0$, $\xi(b) = 0$,

(ii) $\xi(a) = 0$, $\xi(b) \neq 0$,   (iv) $\xi(a) \neq 0$, $\xi(b) \neq 0$.

For the set (i), equation (7.4) gives

$$\delta J = \int_a^b \varepsilon\xi \left\{ \frac{\partial F}{\partial y} - \frac{d}{dx} \frac{\partial F}{\partial y'} \right\} dx, \qquad (7.5)$$

and hence, by the Euler–Lagrange lemma, $\delta J = 0$ implies that y must satisfy

$$\frac{\partial F}{\partial y} - \frac{d}{dx} \frac{\partial F}{\partial y'} = 0, \quad a < x < b. \qquad (7.6)$$

Using this and considering the set (iv), we have

$$\delta J = \left[\varepsilon\xi \frac{\partial F}{\partial y'} \right]_a^b. \qquad (7.7)$$

For this to vanish for arbitrary ξ we must have

$$\frac{\partial F}{\partial y'} = 0 \quad \text{at} \quad x = a \quad \text{and} \quad x = b. \qquad (7.8)$$

Thus the critical curve y must satisfy the Euler–Lagrange equation (7.6) and the boundary conditions (7.8). These boundary conditions for y are called *natural* boundary conditions since they arise naturally in the variational problem. This distinguishes them from the *essential* boundary conditions on y in the fixed end point problem.

EXAMPLE 7.1. Let

$$J(Y) = \int_0^1 (Y')^2 \, dx \qquad (7.9)$$

and consider the problem of finding the critical curves for

(i) fixed end points $Y(0) = 0$, $Y(1) = 1$,
(ii) free end points, $Y(0)$ and $Y(1)$ not prescribed.

The Euler–Lagrange equation is

$$y'' = 0, \qquad (7.10)$$

which has general solution

$$y = \alpha x + \beta \quad (\alpha, \beta \text{ constants}). \qquad (7.11)$$

For (i) the boundary conditions give the critical curve $y = x$, and $J(y) = 1$. For (ii) the natural boundary conditions (7.8), that is

$$\frac{\partial F}{\partial y'} = 2y' = 0 \quad \text{at} \quad x = 0 \quad \text{and} \quad x = 1$$

give the critical curves $y = \beta$, and $J(y) = 0$.

This example illustrates the part played by the boundary conditions in variational problems.

CASE 7.2. *End points variable in x and y directions*

Case 7.1 dealt with end points variable in the y direction. Now we generalise this case and allow the end points to vary in the x and y directions (see Figure 8). This will lead to a formula for the general first variation.

Let

$$J(y) = \int_a^b F(x, y, y') \, dx, \qquad (7.12)$$

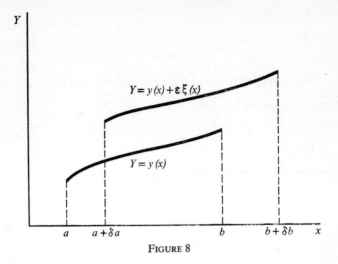

FIGURE 8

and write

$$y(a) = y_a, \quad y(b) = y_b. \tag{7.13}$$

Suppose that the varied curve $Y = y + \varepsilon\xi$ is defined over $(a + \delta a, \, b + \delta b)$ so that

$$J(y + \varepsilon\xi) = \int_{a+\delta a}^{b+\delta b} F(x, \, y + \varepsilon\xi, \, y' + \varepsilon\xi') \, dx, \tag{7.14}$$

and at the end points of this curve we write

$$y + \varepsilon\xi = y_a + \delta y_a \quad \text{at} \quad x = a + \delta a,$$
$$= y_b + \delta y_b \quad \text{at} \quad x = b + \delta b. \tag{7.15}$$

To find the first variation δJ we must obtain the terms in $J(y + \varepsilon\xi)$ which are linear in the first-order quantities $\varepsilon\xi$, δa, δb, δy_a

26

and δy_b. Consider

$$\Delta J = J(y+\varepsilon\xi)-J(y)$$
$$= \int_{a+\delta a}^{b+\delta b} F(x, y+\varepsilon\xi, y'+\varepsilon\xi')\,dx - \int_a^b F(x, y, y')\,dx,$$

which we rewrite as

$$\Delta J = \int_a^b \{F(x, y+\varepsilon\xi, y'+\varepsilon\xi')-F(x, y, y')\}\,dx$$

$$+ \int_b^{b+\delta b} F(x, y+\varepsilon\xi, y'+\varepsilon\xi')\,dx$$

$$- \int_a^{a+\delta a} F(x, y+\varepsilon\xi, y'+\varepsilon\xi')\,dx. \qquad (7.16)$$

We have to assume here that y and $\varepsilon\xi$ are defined in the interval $[a, b+\delta b]$, that is, extensions of y and $y+\varepsilon\xi$ are needed, and we suppose this is done (see equations (7.18) and (7.19) below). Then

$$\Delta J = \int_a^b \left\{\varepsilon\xi\frac{\partial F}{\partial y}+\varepsilon\xi'\frac{\partial F}{\partial y'}+0(\varepsilon^2)\right\}\,dx$$

$$+ \int_b^{b+\delta b} \{F(x, y, y')+0(\varepsilon)\}\,dx$$

$$- \int_a^{a+\delta a} \{F(x, y, y')+0(\varepsilon)\}\,dx$$

and the linear terms in this are

$$\delta J = \int_a^b \left\{\varepsilon\xi\frac{\partial F}{\partial y}+\varepsilon\xi'\frac{\partial F}{\partial y'}\right\}dx + \left[F(x, y, y')\,\delta x\right]_a^b$$

$$= \int_a^b \varepsilon\xi\left\{\frac{\partial F}{\partial y}-\frac{d}{dx}\frac{\partial F}{\partial y'}\right\}dx + \left[\varepsilon\xi\frac{\partial F}{\partial y'}+F(x, y, y')\,\delta x\right]_a^b. \quad (7.17)$$

The final step is to find ξ at $x = a$ and $x = b$. Using (7.15) we have

$$y_a + \delta y_a = y(a + \delta a) + \varepsilon \xi(a + \delta a)$$
$$= y(a) + \delta a\, y'(a) + \ldots + \varepsilon \xi(a) + \delta a \varepsilon \xi'(a) + \ldots,$$

and, since $y_a = y(a)$ by (7.13), this gives

$$\varepsilon \xi(a) = \delta y_a - \delta a\, y'(a) \quad \text{to first order.} \qquad (7.18)$$

Similarly, we find

$$\varepsilon \xi(b) = \delta y_b - \delta b\, y'(b) \quad \text{to first order.} \qquad (7.19)$$

These are the extensions referred to in the text between equations (7.16) and (7.17). With (7.18) and (7.19) we can now write

$$\delta J = \int_a^b \varepsilon \xi \left\{ \frac{\partial F}{\partial y} - \frac{\mathrm{d}}{\mathrm{d}x}\frac{\partial F}{\partial y'} \right\} \mathrm{d}x + \left[\frac{\partial F}{\partial y'} \delta y - \left(y' \frac{\partial F}{\partial y'} - F \right) \delta x \right]_a^b. \qquad (7.20)$$

This is known as the *general first variation*.

If we introduce the canonical variable p and the Hamiltonian H,

$$p = \frac{\partial F}{\partial y'}, \quad H = py' - F, \qquad (7.21)$$

discussed in section 6, we see that δJ becomes

$$\delta J = \int_a^b \varepsilon \xi \left\{ \frac{\partial F}{\partial y} - \frac{\mathrm{d}}{\mathrm{d}x}\frac{\partial F}{\partial y'} \right\} \mathrm{d}x + \left[p\delta y - H\delta x \right]_{x=a}^{x=b}. \qquad (7.22)$$

If the functional $J(Y)$ in (7.12) has an extremum for $Y = y$, then (7.22) shows that y must satisfy the conditions

$$\frac{\partial F}{\partial y} - \frac{\mathrm{d}}{\mathrm{d}x}\frac{\partial F}{\partial y'} = 0, \quad a < x < b, \qquad (7.23)$$

$$\left[p\delta y - H\delta x \right]_{x=a}^{x=b} = 0. \qquad (7.24)$$

28

For fixed end points, $\delta x = 0$ and $\delta y = 0$, and these results reduce to those of section 4.

For end points variable in the y direction only, $\delta x = 0$ and $\delta y = \varepsilon \xi$, and we recover the expression in equation (7.4).

8. Hamilton–Jacobi equation

The general first variation derived above can now be used to obtain an important result known as the Hamilton–Jacobi equation.

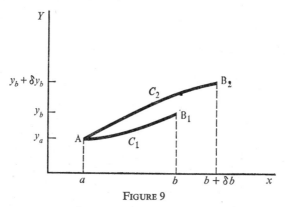

FIGURE 9

We consider the integral

$$J(Y) = \int_a^b F(x, Y, Y') \, dx \qquad (8.1)$$

and suppose that for any two end points $A(a, y_a)$ and $B(b, y_b)$ there is only one critical curve C. Take A fixed and consider two right hand end points

$$\mathbf{B}_1(b, y_b) \quad \text{and} \quad \mathbf{B}_2(b + \delta b, y_b + \delta y_b). \qquad (8.2)$$

The corresponding critical curves are C_1 and C_2 say (Figure 9). Then the integral (8.1) evaluated along any critical curve is just a function of the end points A and B, and since A is fixed, we can regard (8.1) as a function of B alone. Thus

$$J(C_1) = \int_{C_1} F(x, y, y') \, dx \qquad (8.3)$$

is a function S of B_1 which we can write as

$$S = S(b, y_b). \qquad (8.4)$$

Similarly

$$S + \delta S = S(b + \delta b, y_b + \delta y_b) \qquad (8.5)$$

is the corresponding value for the critical curve C_2 joining A and B_2. From these we have

$$\delta S = S(b + \delta b, y_b + \delta y_b) - S(b, y_b), \qquad (8.6)$$

and using the general first variation (7.22) to evaluate the right hand side, we find that to first order

$$\delta S = p \, \delta y_b - H \, \delta b. \qquad (8.7)$$

Hence

$$\frac{\partial S}{\partial y_b} = p, \quad \frac{\partial S}{\partial b} = -H. \qquad (8.8)$$

Now $B_1(b, y_b)$ may be any end point, and so we can replace it by the general point $B(x, y)$ by changing b to x, y_b to y. Then (8.8) becomes

$$\frac{\partial S}{\partial y} = p, \quad \frac{\partial S}{\partial x} = -H, \qquad (8.9)$$

where

$$p = p(x, y) = \frac{\partial F}{\partial y'}, \qquad (8.10)$$

and

$$H = H(x, y, p) = py' - F. \tag{8.11}$$

In (8.10) y' denotes the derivative dy/dx calculated at the point B for the extremal C going from A to B. From (8.9) we have

$$\frac{\partial S}{\partial x} + H\left(x, y, \frac{\partial S}{\partial y}\right) = 0. \tag{8.12}$$

This partial differential equation, which in general is nonlinear, is the *Hamilton–Jacobi equation*.

If the solution S of the Hamilton–Jacobi equation can be found, the critical curves may be obtained directly. To see this suppose that

$$S = S(x, y, \alpha), \quad y = y(x) \quad \text{is a critical curve,} \tag{8.13}$$

is a solution of (8.12), depending on a parameter α. Then consider

$$\frac{d}{dx}\left(\frac{\partial S}{\partial \alpha}\right) = \frac{\partial^2 S}{\partial x\,\partial \alpha} + \frac{\partial^2 S}{\partial y\,\partial \alpha}\frac{dy}{dx}. \tag{8.14}$$

Next, we shall differentiate (8.12) with respect to α. Observe that α occurs only in the third variable of H, which was originally denoted by p. Hence

$$\frac{\partial^2 S}{\partial x\,\partial \alpha} = -\frac{\partial H}{\partial p}\frac{\partial^2 S}{\partial \alpha\,\partial y} \tag{8.15}$$

Putting (8.15) in (8.14) we therefore find

$$\frac{d}{dx}\left(\frac{\partial S}{\partial \alpha}\right) = \frac{\partial^2 S}{\partial y\,\partial \alpha}\left(\frac{dy}{dx} - \frac{\partial H}{\partial p}\right). \tag{8.16}$$

Now, since

$$\frac{dy}{dx} = \frac{\partial H}{\partial p} \quad \text{(canonical equation)}$$

31

along each critical curve, it follows that

$$\frac{d}{dx}\left(\frac{\partial S}{\partial \alpha}\right) = 0$$

or

$$\frac{\partial S}{\partial \alpha} = \text{constant} \quad \text{(on each critical curve)}. \qquad (8.17)$$

We have therefore proved

THEOREM 8.1. Let

$$S = S(x, y, \alpha)$$

be a solution of the Hamilton–Jacobi equation (8.12) depending on a parameter α (constant of integration). Then

$$\frac{\partial S}{\partial \alpha} = \text{constant}$$

along each critical curve.

EXAMPLE 8.1. To illustrate this result we take the integral

$$J(y) = \int_a^b y'^2 \, dx. \qquad (8.18)$$

Here $F(x, y, y') = y'^2$, $p = \partial F/\partial y' = 2y'$. So the Hamiltonian is given by

$$H(x, y, p) = py' - F = \tfrac{1}{2}p^2 - \tfrac{1}{4}p^2 = \tfrac{1}{4}p^2.$$

The Hamilton–Jacobi equation (8.12)

$$\frac{\partial S}{\partial x} + H\left(x, y, \frac{\partial S}{\partial y}\right) = 0$$

becomes

$$\frac{\partial S}{\partial x} + \tfrac{1}{4}\left(\frac{\partial S}{\partial y}\right)^2 = 0. \tag{8.19}$$

This is a first order nonlinear partial differential equation. To solve it for S we take

$$S = S(x, y) = u(x) + v(y), \tag{8.20}$$

which gives

$$\frac{du}{dx} + \tfrac{1}{4}\left(\frac{dv}{dy}\right)^2 = 0. \tag{8.21}$$

It follows from (8.21) that du/dx must be constant, because du/dx does not depend on y and $(dv/dy)^2$ does not depend on x. Hence

$$u = -\alpha^2 x \quad (\alpha = \text{const.}).$$

Then

$$-\alpha^2 + \tfrac{1}{4}\left(\frac{dv}{dy}\right)^2 = 0,$$

which gives

$$\frac{dv}{dy} = 2\alpha,$$

or

$$v = 2\alpha y + \beta,$$

where β is another constant. So, by (8.20),

$$S = -\alpha^2 x + 2\alpha y + \beta. \tag{8.22}$$

From Theorem 8.1, the critical curves are given by

$$\frac{\partial S}{\partial \alpha} = \text{const.}, \quad \text{i.e. } y = \alpha x + c \quad (\alpha, c \text{ constants}). \tag{8.23}$$

33

$\partial S/\partial \beta$ = constant is an identity and so does not give anything new. The critical curves in (8.23) are straight lines, in agreement with the result in (7.11) obtained by solving the Euler–Lagrange equation for the integral (8.18).

9. Variational principles of mechanics

9.1. HAMILTON'S PRINCIPLE

We now turn to some variational aspects of classical mechanics, and first we consider a particle of mass m and position $\mathbf{r}$ moving in three dimensional space (x, y, z) under the action of a force $\mathbf{F}$ which is conservative so that $\mathbf{F} = -\operatorname{grad} V$, where $V = V(x, y, z)$ is the potential. Then Newton's law of motion states that

$$m \frac{\mathrm{d}^2 \mathbf{r}}{\mathrm{d}t^2} = \mathbf{F}, \tag{9.1}$$

or

$$m\ddot{x} + \frac{\partial V}{\partial x} = 0, \quad m\ddot{y} + \frac{\partial V}{\partial y} = 0, \quad m\ddot{z} + \frac{\partial V}{\partial z} = 0. \tag{9.2}$$

Consider the first equation in (9.2). It may be written as

$$\frac{\mathrm{d}}{\mathrm{d}t} \frac{\partial}{\partial \dot{x}} \left(\tfrac{1}{2} m\dot{x}^2 \right) + \frac{\partial V}{\partial x} = 0, \tag{9.3}$$

and since $V = V(x, y, z)$ here, we can rewrite this as

$$\frac{\partial L}{\partial x} - \frac{\mathrm{d}}{\mathrm{d}t} \frac{\partial L}{\partial \dot{x}} = 0, \tag{9.4}$$

where

$$L = T - V, \tag{9.5}$$

T being the kinetic energy

$$T = \tfrac{1}{2} m\dot{\mathbf{r}}^2 = \tfrac{1}{2} m(\dot{x}^2 + \dot{y}^2 + \dot{z}^2). \tag{9.6}$$

34

In a similar way, the remaining equations in (9.2) become

$$\frac{\partial L}{\partial y} - \frac{\mathrm{d}}{\mathrm{d}t}\frac{\partial L}{\partial \dot{y}} = 0, \quad \frac{\partial L}{\partial z} - \frac{\mathrm{d}}{\mathrm{d}t}\frac{\partial L}{\partial \dot{z}} = 0, \qquad (9.7)$$

and from our earlier work we recognise equations (9.4) and (9.7) as the Euler–Lagrange equations for the integral

$$J(x, y, z) = \int_{t_1}^{t_2} L(t, x, y, z, \dot{x}, \dot{y}, \dot{z}) \, \mathrm{d}t. \qquad (9.8)$$

We have therefore made a connection between Newton's law of motion (9.1) and the critical curves of the integral (9.8). This result is due to Hamilton (1805–65) and is known as:

Hamilton's principle. If a particle moves from a point A to a point B in a time interval $t_1 \leqslant t \leqslant t_2$, then the path it follows is one which makes the integral (9.8) stationary, that is

$$\delta J = \delta \int L \, \mathrm{d}t = 0.$$

The above discussion can be extended to cover a system of many particles, and other coordinate systems. If the system can be described by generalised coordinates $q_1, \ldots, q_n$ with kinetic energy

$$T = T(q_1, \ldots, q_n, \dot{q}_1, \ldots, \dot{q}_n) \qquad (9.9)$$

and potential energy

$$V = V(q_1, \ldots, q_n), \qquad (9.10)$$

Hamilton's principle holds for the integral

$$J(q_1, \ldots, q_n) = \int_{t_1}^{t_2} L(t, q_1, \ldots, q_n, \dot{q}_1, \ldots, \dot{q}_n) \, \mathrm{d}t, \qquad (9.11)$$

where L is the Lagrangian given by

$$L = T - V. \tag{9.12}$$

The critical curves of $J(q_1, \ldots, q_n)$ are solutions of the equations

$$J_i' \equiv \frac{\partial L}{\partial q_i} - \frac{\mathrm{d}}{\mathrm{d}t} \frac{\mathrm{d}L}{\mathrm{d}\dot{q}_i} = 0, \quad i = 1, \ldots, n, \tag{9.13}$$

which, in classical mechanics, are known as *Lagrange's equations*. These general equations can readily be shown[†] to be equivalent to Newton's equations of motion. Equations (9.4) and (9.7) are examples of Lagrange's equations.

Thus, Newton's equations, Lagrange's equations, and Hamilton's variational principle are equivalent ways of expressing the physical laws of motion in classical dynamics.

9.2. HAMILTON'S PRINCIPLE IN CANONICAL FORM

The results of section 6 can be used to give a canonical form of Hamilton's principle. As in equation (6.3) we can define the variables

$$p_i = \frac{\partial L}{\partial \dot{q}_i}, \quad i = 1, \ldots, n, \tag{9.14}$$

which are the generalised momenta conjugate to q_i, and then define the Hamiltonian H, as in equation (6.4) by

$$H(t, q_1, \ldots, q_n, p_1, \ldots, p_n) = \sum_{i=1}^{n} p_i \dot{q}_i - L. \tag{9.15}$$

Instead of (9.11) we introduce the functional

$$I(q_1, \ldots, q_n, p_1, \ldots, p_n) = \int_{t_1}^{t_2} \left\{ \sum_{i=1}^{n} p_i \dot{q}_i - H \right\} \mathrm{d}t \tag{9.16}$$

[†] See, for example, H. Goldstein, *Classical Mechanics*, Addison-Wesley, Reading, Mass. (1950).

and we find that the stationary condition

$$\delta I = 0 \qquad (9.17)$$

implies the canonical Euler equations

$$\frac{dq_i}{dt} = \frac{\partial H}{\partial p_i}, \quad -\frac{dp_i}{dt} = \frac{\partial H}{\partial q_i}, \quad i = 1, \ldots, n. \qquad (9.18)$$

These $2n$ equations are known as Hamilton's equations in mechanics, and $\delta I = 0$ in (9.17) is a canonical form of Hamilton's variational principle.

9.3. Principle of Stationary Action

Now we consider a more general variation of the path of the system. This will allow us to vary the time of transit from one state $\mathbf{q} = \boldsymbol{\beta}$ to another state $\mathbf{q} = \boldsymbol{\gamma}$.

To do this we use the formula (7.20) for the general first variation, which gives for the integral (9.11) and the variation in Figure 10

$$\delta J = \int_{t_1}^{t_2} \sum_{i=1}^{n} \varepsilon \xi_i \left\{ \frac{\partial L}{\partial q_i} - \frac{d}{dt} \frac{\partial L}{\partial \dot{q}_i} \right\} dt - \left[\left(\sum_{i=1}^{n} \dot{q}_i \frac{\partial L}{\partial \dot{q}_i} - L \right) \delta t \right]_{t=t_1}^{t=t_2}.$$

$$(9.19)$$

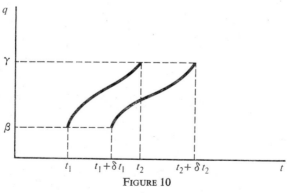

FIGURE 10

37

If the q_i satisfy Lagrange's equations of motion (9.13), we have

$$\delta J = -\left[\left(\sum_{i=1}^{n} \dot{q}_i \frac{\partial L}{\partial \dot{q}_i} - L\right)\delta t\right]_{t=t_1}^{t=t_2}$$

$$= -[H\,\delta t]_{t=t_1}^{t=t_2} \tag{9.20}$$

by (9.14) and (9.15). Now suppose that H is equal to the *same constant* on all admissible paths. Then

$$\delta J = -H(\delta t_2 - \delta t_1) = -H\delta \int_{t_1}^{t_2} \mathrm{d}t$$

$$= -\delta \int_{t_1}^{t_2} H\,\mathrm{d}t. \tag{9.21}$$

But from the canonical form of J given by (9.16) we have

$$\delta J = \delta \int_{t_1}^{t_2} \left(\sum_{i=1}^{n} p_i \dot{q}_i - H\right) \mathrm{d}t$$

and so (9.21) becomes

$$\delta \int_{t_1}^{t_2} \sum_{i=1}^{n} p_i \dot{q}_i \, \mathrm{d}t = 0. \tag{9.22}$$

This is called the *principle of stationary action*. It is not as useful as Hamilton's principle because it is restricted to the case where H is equal to the same constant on all admissible paths. The interest in this principle is that it is essentially the original variational principle of Maupertuis (1698–1759) and the earlier principle of Leibniz (1646–1716).

10. Exercises

1. Find the critical curves of the functional

$$J(y, z) = \int_0^{\pi/2} (y'^2 + z'^2 + 2yz) \, \mathrm{d}x$$

subject to the boundary conditions

$$y(0) = 0, \quad y(\pi/2) = 1, \quad z(0) = 0, \quad z(\pi/2) = 1.$$

2. Show that the functionals

$$\int_a^b F(x, y_1, \ldots, y_n, y_1', \ldots, y_n') \, \mathrm{d}x$$

and

$$\int_a^b \{F(x, y_1, \ldots, y_n, y_1', \ldots, y_n') + \Psi(x, y_1, \ldots, y_n, y_1', \ldots, y_n')\} \, \mathrm{d}x$$

lead to the same Euler–Lagrange equations if

$$\Psi = \frac{\partial \Phi}{\partial x} + \sum_{i=1}^n \frac{\partial \Phi}{\partial y_i} y_i',$$

where $\Phi = \Phi(x, y_1, \ldots, y_n)$ is any twice differentiable continuous function.

3. Solve the Hamilton–Jacobi equation corresponding to the functional

$$J(y) = \int_{x_0}^{x_1} f(y) (1 + y'^2)^{1/2} \, \mathrm{d}x$$

and use the result to find the critical curves of $J(y)$.

4. Find a functional which leads to the Hamilton–Jacobi equation

$$\left(\frac{\partial S}{\partial x}\right)^2 + \left(\frac{\partial S}{\partial y}\right)^2 = 1.$$

5. The functional

$$J(x) = \tfrac{1}{2} \int_{t_1}^{t_2} (m\dot{x}^2 - kx^2)\, dt = \int_{t_1}^{t_2} L(t, x, \dot{x})\, dt$$

corresponds to a simple harmonic oscillator, that is, a particle of mass m acted on by a restoring force $-kx$. Obtain the Lagrange equation of motion, and the Hamilton (canonical) equations of motion.

6. The functional

$$J(r, \theta) = \int_{t_1}^{t_2} \left\{ \tfrac{1}{2} m \, (\dot{r}^2 + r^2\dot{\theta}^2) + \frac{k}{r} \right\} dt = \int_{t_1}^{t_2} L(t, r, \theta, \dot{r}, \dot{\theta})\, dt$$

corresponds to the plane motion of a particle of mass m attracted to the origin of coordinates by a force $-k/r^2$, where k is a constant and r, θ are the usual polar coordinates. Find the Lagrange equations, the Hamiltonian H, and the Hamilton equations of motion.

CHAPTER THREE

Minimum Principles

---

11. Introduction

We have seen in section 3 that a necessary condition for $J(Y)$ to have a minimum at $Y = y$, that is

$$J(y) \leqslant J(Y) \quad \text{for all} \quad Y \in \Omega, \tag{11.1}$$

is that the stationary condition

$$\delta J = \langle \varepsilon\xi, J'(y) \rangle = 0 \tag{11.2}$$

holds for all admissible $\varepsilon\xi$. Various consequences of this stationary condition have been worked out in subsequent sections, but of course the stationary result does not tell us whether $J(Y)$ has a maximum, or a minimum, or neither, at $Y = y$. Second and higher order terms in ε must be considered to establish an extremum, and this is what we now look at.

In some variational problems the question of a minimum can be settled directly by looking at the difference

$$\Delta J = J(Y) - J(y) \tag{11.3}$$

and showing that $\Delta J \geqslant 0$ for all $Y \in \Omega$.

For example, consider

$$J(Y) = \int_0^1 \{(Y')^2 + Y^2\}\, dx, \quad Y(0) = 0, \quad Y(1) = 1. \tag{11.4}$$

Then the critical curve is readily found to be

$$y = \sinh x/\sinh 1, \qquad (11.5)$$

and expanding about this function we have

$$\Delta J = J(y+\varepsilon\xi) - J(y) = \varepsilon^2 \int_0^1 \{(\xi')^2 + \xi^2\} \, dx$$

$$\geqslant 0 \quad \text{for all } \xi. \qquad (11.6)$$

Thus the function (11.5) does minimise $J(Y)$ in (11.4).

As a second example, consider

$$J(Y) = \int_0^1 \{(Y')^2 + (Y')^3\} \, dx, \quad Y(0) = 0 = Y(1). \quad (11.7)$$

This has critical curve $y = 0$, and expanding about this function we have

$$\Delta J = J(y+\varepsilon\xi) - J(y) = \varepsilon^2 \int_0^1 \xi'^2 \, dx + \varepsilon^3 \int_0^1 \xi'^3 \, dx. \quad (11.8)$$

Now the sign of ΔJ in (11.8) is not definite and so $y = 0$ does not necessarily make $J(y)$ a minimum. On the other hand, if we *restrict* the admissible curves to those which have $\varepsilon\xi'$ sufficiently small, so that

$$\Delta J = \int_0^1 \varepsilon^2\xi'^2(1 + \varepsilon\xi') \, dx$$

$$\geqslant 0, \qquad (11.9)$$

that is, we require

$$\varepsilon\xi' > -1, \qquad (11.10)$$

then we do get a minimum for J at $Y = y = 0$. This example shows that more specific statements are needed about the properties of the admissible curves.

42

The two cases corresponding to

ξ' restricted   (weak variations)   (11.11)

ξ' unrestricted   (strong variations)   (11.12)

are dealt with in more advanced texts. The usual approach is to study second order terms in $J(y+\varepsilon\xi)$, by analogy with the ordinary calculus case described in section 3. Here we shall confine our attention to simple cases which do not require elaborate treatment.

12. Quadratic problems

An interesting class of variational problems involves the quadratic functional

$$J(Y) = \tfrac{1}{2} \int_a^b \{(Y')^2\, v + wY^2 - 2rY\}\ \mathrm{d}x, \qquad (12.1)$$

with

$$Y(a) = y_a, \quad Y(b) = y_b, \qquad (12.2)$$

where v, w and r are prescribed functions of x in general. The corresponding Euler–Lagrange equation is

$$-\frac{\mathrm{d}}{\mathrm{d}x}\left\{ v\,\frac{\mathrm{d}y}{\mathrm{d}x} \right\} + wy = r, \qquad a < x < b, \qquad (12.3)$$

with

$$y(a) = y_a, \quad y(b) = y_b, \qquad (12.4)$$

involving a Sturm–Liouville equation (12.3). If y denotes the critical curve satisfying (12.3) and (12.4), then (12.1) gives

$$\Delta J = J(y+\varepsilon\xi) - J(y) = \tfrac{1}{2} \int_a^b \{v(\varepsilon\xi')^2 + w(\varepsilon\xi)^2\}\ \mathrm{d}x, \quad (12.5)$$

43

and we want to investigate this for a definite sign property which will lead to an extremum principle.

Suppose for simplicity that v and w are constants, and let v be positive (if v is negative change the sign of J). Then

$$\Delta J = \tfrac{1}{2}\, v\varepsilon^2 \int_a^b \left\{ (\xi')^2 + \frac{w}{v}\, \xi^2 \right\} dx \qquad (12.6)$$

and we look for a result which ensures that the integral in (12.6) is of one sign. So consider

$$K(\xi) = \int_a^b \left\{ (\xi')^2 + \frac{w}{v}\, \xi^2 \right\} dx, \quad \xi(a) = \xi(b) = 0. \quad (12.7)$$

Integrate the ξ' term by parts, to obtain

$$K(\xi) = \int_a^b \xi \left\{ -\frac{d^2}{dx^2} + \frac{w}{v} \right\} \xi \, dx. \qquad (12.8)$$

Now $\xi(x)$ is zero at $x = a$ and $x = b$, and we can expand it in a series* of eigenfunctions ϕ_n of $-d^2/dx^2$ with the same boundary conditions,

$$\xi = \sum_{n=1}^{\infty} a_n \phi_n, \ \phi_n = \sin\left\{ \frac{n\pi(x-a)}{b-a} \right\}. \qquad (12.9)$$

Then (12.8) becomes

$$K(\xi) = \int_a^b \sum_{m=1}^{\infty} a_m \phi_m \sum_{n=1}^{\infty} \left\{ \frac{n^2\pi^2}{(b-a)^2} + \frac{w}{v} \right\} a_n \phi_n \, dx \qquad (12.10)$$

and since

$$\int_a^b \phi_n \phi_m \, dx = 0 \quad \text{for} \quad n \neq m, \qquad (12.11)$$

---

* See, for example, I. N. Sneddon, *Fourier Series*, Routledge & Kegan Paul (Library of Mathematics), London (1961).

we have

$$K(\xi) = \sum_{n=1}^{\infty} a_n^2 \left\{ \frac{n^2\pi^2}{(b-a)^2} + \frac{w}{v} \right\} \int_a^b \phi_n^2 \, dx. \qquad (12.12)$$

It therefore follows from (12.12) that if

$$\frac{\pi^2}{(b-a)^2} + \frac{w}{v} \geqslant 0, \qquad (12.13)$$

then

$$K(\xi) \geqslant 0, \qquad (12.14)$$

which in turn means that

$$\Delta J \geqslant 0, \qquad (12.15)$$

and hence we have a minimum principle.

If $w > 0$, then since $v > 0$, ΔJ in (12.6) is non-negative, but (12.13) allows w to be negative, since

$$w \geqslant -\frac{v\pi^2}{(b-a)^2} \qquad (12.16)$$

is all that is required.

The result in (12.7), (12.13) and (12.14) is a useful inequality which we state as

THEOREM 12.1

$$\int_a^b (u')^2 \, dx \geqslant \frac{\pi^2}{(b-a)^2} \int_a^b u^2 \, dx \qquad (12.17)$$

for all u such that $u(a) = u(b) = 0$.

13. Dynamic programming

At this point we leave the classical Euler–Lagrange theory for a moment and turn to a twentieth-century approach to variational problems known as dynamic programming. This name was introduced by Bellman in 1957 and describes an approach which links up with the Hamilton–Jacobi theory of section 8.

Dynamic programming is a powerful optimisation method which may be applied to any problem whose solution involves a

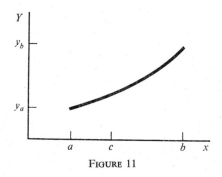

FIGURE 11

multistage decision process. To see what this is we consider the fundamental variational problem for

$$J(Y) = \int_a^b F(x, Y, Y') \, dx, \quad Y(a) = y_a, \quad Y(b) = y_b. \quad (13.1)$$

Let c be some point in (a, b). Suppose we follow the critical curve y along from (a, y_a) and reach the point (c, y). Then, for the rest of the curve on (c, b) to be optimal, say minimising $J(Y)$ on (a, b), we must minimise the integral $\int_c^b F \, dx$. This

46

applies to all the points c in (a, b), and in this example we have a multistage decision process where the number of stages (corresponding to c) increases to infinity.

This idea is formulated as the:

Principle of Optimality. An optimal sequence of decisions in a multistage decision process problem has the property that whatever the initial stage, state and decision are, the remaining decisions must constitute an optimal sequence of decisions for the remaining problem, with the stage and state resulting from the first decision considered as initial conditions.

For our purpose we interpret state to mean the point (x, y) and stage to mean the arc on (a, c).

We now use this principle to discuss the variational problem for

$$J(\Phi) = \int_a^x F(\bar{x}, \Phi, \Phi')\, d\bar{x}, \quad \Phi' \equiv \frac{d\Phi}{d\bar{x}}, \qquad (13.2)$$

$$\Phi(a) = y_a, \qquad\qquad \Phi(x) = y. \qquad (13.3)$$

Note the right hand end point is (x, y), and we use Φ instead of Y to avoid confusion.

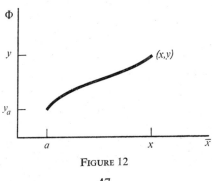

FIGURE 12

We suppose that for given x and y there is some curve $\Phi = \phi$ which minimises $J(\Phi)$. We can define the minimum as

$$S(x, y) = \min_{\Phi} \left\{ \int_a^x F(\bar{x}, \Phi, \Phi') \, d\bar{x} \right\}, \qquad (13.4)$$

which is just the function $S(x, y)$ of equation (8.4).

To use the principle of optimality we divide the interval $[a, x]$ into two parts:

$$[a, x] = [a, x - \Delta x] + [x - \Delta x, x]. \qquad (13.5)$$

On $[a, x - \Delta x]$ we take Φ to be optimal, while on $[x - \Delta x, x]$ Φ is arbitrary except for the end value $\Phi(x) = y$. Then

$$\int_a^x F(\bar{x}, \Phi, \Phi') \, d\bar{x} = \int_a^{x - \Delta x} F(\bar{x}, \Phi, \Phi') \, d\bar{x}$$

$$+ \int_{x - \Delta x}^x F(\bar{x}, \Phi, \Phi') \, d\bar{x}$$

$$= J_1 + J_2, \quad \text{say}, \qquad (13.6)$$

and so

$$S(x, y) \leqslant J_1 + J_2. \qquad (13.7)$$

Now, by (13.4) and the fact that Φ is optimal over $[a, x - \Delta x]$,

$$J_1 = S(x - \Delta x, \Phi(x - \Delta x)) = S(x - \Delta x, y - y'\Delta x + 0(\Delta x^2)), \quad (13.8)$$

where

$$y = \Phi(x) \quad \text{and} \quad y' \overset{\text{defn.}}{=} \Phi'(x). \qquad (13.9)$$

Also,

$$J_2 = F(x, y, y') \, \Delta x + 0(\Delta x^2), \qquad (13.10)$$

where we have used (13.9). In (13.8) and (13.10) the function y' is arbitrary, being Φ' at x. From (13.7), (13.8) and (13.10) we have

$$S(x, y) \leqslant F(x, y, y')\, \Delta x + 0(\Delta x^2) + S(x - \Delta x, y - y'\, \Delta x + 0(\Delta x^2)).$$

Minimising the right hand side over y' gives

$$S(x, y) = \min_{y'} \{ F(x, y, y')\, \Delta x + 0(\Delta x^2)$$
$$+ S(x - \Delta x, y - y'\, \Delta x + 0(\Delta x^2)) \}$$

and expanding the last term we have

$$S(x, y) = \min_{y'} \left\{ F(x, y, y')\, \Delta x \right.$$
$$\left. + S(x, y) - \frac{\partial S}{\partial x}\, \Delta x - \frac{\partial S}{\partial y}\, y'\, \Delta x + 0(\Delta x^2) \right\}.$$

As $\Delta x \to 0$ this gives

$$0 = \min_{y'} \left\{ F(x, y, y') - \frac{\partial S}{\partial x} - y' \frac{\partial S}{\partial y} \right\}. \qquad (13.11)$$

This is the basic partial differential equation of dynamic programming. It was derived earlier, in 1935, from different considerations by Carathéodory (1873–1950).

If the minimum in (13.11) occurs at $y' = \varrho$, we have

$$F = \frac{\partial S}{\partial x} + y' \frac{\partial S}{\partial y} \quad \text{at} \quad y' = \varrho, \qquad (13.12)$$

and

$$\frac{\partial F}{\partial y'} = \frac{\partial S}{\partial y} \quad \text{at} \quad y' = \varrho. \qquad (13.13)$$

See equations (8.9) and (8.10).

Now $p = \partial F / \partial y'$ and so (13.12) and (13.13) give

$$F = \frac{\partial S}{\partial x} + y' \frac{\partial F}{\partial y'},$$

49

or

$$\frac{\partial S}{\partial x} + H\left(x, y, \frac{\partial S}{\partial y}\right) = 0, \qquad (13.14)$$

where H is the Hamiltonian. Equation (13.14) is the Hamilton–Jacobi equation derived in section 8 by a different method.

14. Isoperimetric problems

In example 1.4 we described one of the earliest variational problems which concerns the area enclosed between the curve Y and an interval of the x axis (see Figure 3). The problem was to find the function $Y = y$ which makes the area

$$J(Y) = \int_0^a Y \, dx, \quad Y(0) = Y(a) = 0, \qquad (14.1)$$

a maximum, subject to the length of the curve being fixed, that is

$$K(Y) = \int_0^a \sqrt{(1+Y'^2)} \, dx = \text{constant} = k. \qquad (14.2)$$

This isoperimetric problem is an example of the more general problem of finding the function $Y = y$ which makes

$$J(Y) = \int_a^b F(x, Y, Y') \, dx, \quad Y(a) = y_a, \quad Y(b) = y_b, \qquad (14.3)$$

an extremum, subject to the integral constraint

$$K(Y) = \int_a^b G(x, Y, Y') \, dx = \text{constant}. \qquad (14.4)$$

It was Euler who solved the problem in (14.3) and (14.4). Suppose $Y = y(x)$ is the critical curve and consider admissible curves of the form

$$Y(x) = y(x) + \varepsilon_1 \xi(x) + \varepsilon_2 \eta(x), \qquad (14.5)$$

where ξ and η vanish at $x = a$ and $x = b$, so that $Y(x)$ satisfies the boundary conditions in (14.3). Then

$$J(y + \varepsilon_1 \xi + \varepsilon_2 \eta) = J(y) + \langle \varepsilon_1 \xi + \varepsilon_2 \eta, J'(y) \rangle + 0(\varepsilon^2), \quad (14.6)$$

and

$$K(y + \varepsilon_1 \xi + \varepsilon_2 \eta) = K(y) + \langle \varepsilon_1 \xi + \varepsilon_2 \eta, K'(y) \rangle + 0(\varepsilon^2). \quad (14.7)$$

Using the method of Lagrange multipliers we then set

$$\frac{\partial}{\partial \varepsilon_1} \left\{ \lambda_0 J(y + \varepsilon_1 \xi + \varepsilon_2 \eta) \right.$$

$$\left. + \lambda_1 K(y + \varepsilon_1 \xi + \varepsilon_2 \eta) \right\} = 0 \quad \text{for} \quad \varepsilon_1 = \varepsilon_2 = 0, \quad (14.8)$$

and

$$\frac{\partial}{\partial \varepsilon_2} \left\{ \lambda_0 J(y + \varepsilon_1 \xi + \varepsilon_2 \eta) \right.$$

$$\left. + \lambda_1 K(y + \varepsilon_1 \xi + \varepsilon_2 \eta) \right\} = 0 \quad \text{for} \quad \varepsilon_1 = \varepsilon_2 = 0, \quad (14.9)$$

where λ_0 and λ_1 are unknown parameters at this stage. By (14.6) and (14.7) these give

$$\lambda_0 \langle \xi, J'(y) \rangle + \lambda_1 \langle \xi, K'(y) \rangle = 0 \qquad (14.10)$$

and

$$\lambda_0 \langle \eta, J'(y) \rangle + \lambda_1 \langle \eta, K'(y) \rangle = 0. \qquad (14.11)$$

Now (14.10) implies that λ_0/λ_1 is independent of η. Since η is arbitrary in (a, b), equation (14.11) gives

$$\lambda_0 J'(y) + \lambda_1 K'(y) = 0. \qquad (14.12)$$

If we assume that y is *not* a critical curve of $K(Y)$, that is $K'(y) \neq 0$, equation (14.12) means that y is a solution of

$$J'(y) + \lambda K'(y) = 0 \quad (\lambda = \lambda_1/\lambda_0), \tag{14.13}$$

that is,

$$\frac{\partial}{\partial y}(F + \lambda G) - \frac{d}{dx}\frac{\partial}{\partial y'}(F + \lambda G) = 0. \tag{14.14}$$

This is *Euler's rule*. Equation (14.14) is in general a second order differential equation and so will have two constants of integration which, together with the unknown parameter λ, are determined by the boundary conditions $y(a) = y_a, y(b) = y_b$ and the constraint (14.4).

What we have derived here is, of course, just the stationary condition. The extremum property is usually clear from the context of the problem.

EXAMPLE 14.1. We now apply Euler's rule to the isoperimetric problem in equations (14.1) and (14.2). For that case

$$F = Y \quad \text{and} \quad G = \sqrt{(1 + Y'^2)}. \tag{14.15}$$

According to (14.14) we require the Euler equation for

$$F + \lambda G = Y + \lambda\sqrt{(1 + Y'^2)}. \tag{14.16}$$

Setting this in (14.14) we obtain

$$1 - \lambda \frac{d}{dx}\{y'(1 + y'^2)^{-1/2}\} = 0.$$

Integrating once yields

$$x - \lambda y'(1 + y'^2)^{-1/2} = C_1,$$

where C_1 is a constant of integration. On solving for y' and integrating again we find that

$$(x - C_1)^2 + (y - C_2)^2 = \lambda^2, \tag{14.17}$$

where C_2 is another constant of integration. The curves (14.17) are arcs of circles with centre (C_1, C_2) and radius λ. The values of C_1, C_2 and λ are determined by the conditions $y(0) = y(a) = = 0$, and

$$\int_0^a (1+y'^2)^{1/2} \, dx = k.$$

EXAMPLE 14.2. *An eigenvalue problem*

Another interesting constraint problem arises when we want to minimise

$$J(Y) = \int_0^1 (Y')^2 \, dx, \quad Y(0) = Y(1) = 0, \quad (14.18)$$

subject to

$$K(Y) = \int_0^1 Y^2 \, dx = 1. \quad (14.19)$$

The Euler equation (14.14) for this problem is

$$y'' - \lambda y = 0, \quad 0 < x < 1. \quad (14.20)$$

To satisfy the boundary conditions we have

$$y = A \sin n\pi x, \quad n = 1, 2, \ldots, \quad (14.21)$$

where

$$-\lambda = n^2\pi^2, \quad n = 1, 2, \ldots . \quad (14.22)$$

The multiplier λ is playing the part of an eigenvalue in (14.20). For the functions (14.21), the constraint (14.19) requires $A = \sqrt{2}$. Then

$$J(y) = A^2 \int_0^1 n^2\pi^2 \cos^2 n\pi x \, dx = n^2\pi^2, \quad n = 1, 2, \ldots . \quad (14.23)$$

The lowest value of $J(y)$ is therefore

$$J(y) = \pi^2, \qquad (14.24)$$

corresponding to $n = 1$ and $y = \sqrt{2} \sin \pi x$.

15. Exercises

1. Find the curve joining the points $(-1, 1)$ and $(1, 1)$ which minimises the functional

$$J(y) = \int_{-1}^{1} (x^2 y'^2 + 12 y^2) \, dx.$$

2. Find the curve joining the points $(1, 3)$ and $(2, 5)$ which minimises the functional

$$J(y) = \int_{1}^{2} y'(1 + x^2 y') \, dx.$$

3. Show that $y = \sin 2x - 1$ provides a maximum (strong) for the functional

$$J(y) = \int_{0}^{\pi/4} (4y^2 - y'^2 + 8y) \, dx, \quad y(0) = -1, \quad y\left(\frac{\pi}{4}\right) = 0.$$

4. Find the curve which minimises the functional

$$J(y) = \int_{0}^{1} (y'^2 + y^2 + 2ye^{2x}) \, dx,$$

with $y(0) = \frac{1}{3}$, $y(1) = \frac{1}{3}e^2$.

5. Find the critical curves of the isoperimetric problem

$$J(y) = \int_{0}^{1} (y'^2 + x^2) \, dx, \quad y(0) = y(1) = 0,$$

with

$$\int_0^1 y^2 \, dx = 2.$$

6. Find the function which minimises the functional

$$J(y) = \int_0^1 y'^2 \, dx, \quad y(0) = 0, \quad y(1) = 1,$$

with

$$\int_0^1 y \, dx = \tfrac{1}{3}.$$

7. Find the function y which maximises the integral

$$J(y) = -\int_{-\infty}^{\infty} y \log \, y \, dx \quad (y \geqslant 0)$$

subject to two constraints

$$\int_{-\infty}^{\infty} y \, dx = 1, \quad \int_{-\infty}^{\infty} x^2 y \, dx = \sigma^2,$$

where $y \to 0$ as $x \to \pm \infty$. This problem arises in the theory of probability where y is a probability density and $J(y)$ is a quantity known as the entropy.

8. Investigate the isoperimetric problem for

$$J(y) = \int_0^1 y'^2 \, dx$$

subject to the conditions

$$y(0) = 0, \quad y(1) = 1, \quad K(y) = \int_0^1 (1 + y'^2)^{1/2} \, dx = 2.$$

55

CHAPTER FOUR

Direct Methods

---

16. Rayleigh–Ritz method

So far in this book we have dealt with variational problems by reducing them to problems in differential equations (the Euler–Lagrange theory), which were then solved for the critical curves. However, for many problems the Euler–Lagrange equations are not solvable exactly and this classical procedure is therefore not very useful. For example, the problem for

$$J(Y) = \int_0^1 \{\tfrac{1}{2}(Y')^2 + e^Y\} \, dx, \quad Y(0) = 0, \quad Y(1) = 1,$$

(16.1)

has Euler–Lagrange equation

$$y'' = e^y, \quad 0 < x < 1$$

(16.2)

with

$$y(0) = 0, \quad y(1) = 1,$$

(16.3)

and this nonlinear differential equation cannot be solved exactly. This situation has led to the development of *direct variational methods* in which one works directly with the functional $J(Y)$, rather than indirectly via the Euler–Lagrange differential equation.

One of the most widely used direct methods is due to Rayleigh and Ritz and this is the one which we shall now describe.

We shall explain the method through a simple quadratic variational problem, though the ideas have wide applicability.

Consider the problem for

$$J(Y) = \int_0^1 \{ \tfrac{1}{2}(Y')^2 + \tfrac{1}{2}wY^2 - qY \} \, dx, \quad Y(0) = Y(1) = 0,$$

$$(16.4)$$

where w is non-negative. Then, by the results of section 12, we know that this integral has a minimum for the critical curve y which satisfies

$$\frac{-d^2y}{dx^2} + wy = q, \quad 0 < x < 1, \tag{16.5}$$

$$y(0) = y(1) = 0. \tag{16.6}$$

We shall bypass the Euler–Lagrange equation (16.5) and concentrate on the minimum principle

$$J(y) \leqslant J(Y), \quad Y \in \Omega, \tag{16.7}$$

that is

$$J(y) = \min_{Y \in \Omega} J(Y). \tag{16.8}$$

In the Rayleigh–Ritz method we minimise $J(Y)$ not for the complete space Ω of admissible functions, but for a finite space R_n of functions. Then we have

$$J(y) = \min_{Y \in \Omega} J(Y) \leqslant \min_{Y \in R_n} J(Y). \tag{16.9}$$

The idea therefore is to evaluate the right hand side of (16.9) exactly, and hence obtain an upper bound for $J(y)$. At the same time, the minimising function Y in R_n is an approximation to the exact function y.

For example, we could start with $n = 1$ and take

$$Y_1 = \alpha_1 x(1 - x), \tag{16.10}$$

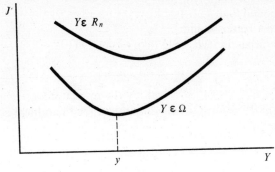

FIGURE 13

which satisfies the exact boundary conditions in (16.4), where α_1 is a parameter. We put this in $J(Y)$ which gives a function J of α_1. We then minimise $J(\alpha_1)$ by solving

$$\frac{\mathrm{d}J}{\mathrm{d}\alpha_1} = 0 \qquad (16.11)$$

for α_1. This optimum value of α_1 is put back into $J(\alpha_1)$ and hence we obtain an upper bound for $J(y)$

$$J(y) \leqslant \min_{Y_1} J(Y_1). \qquad (16.12)$$

This process can be continued, by taking for example

$$Y_2 = \alpha_1 x(1-x) + \alpha_2 x^2(1-x)^2 \qquad (16.13)$$

and finding the optimum values of α_1 and α_2 from

$$\frac{\partial J}{\partial \alpha_1} = 0, \qquad \frac{\partial J}{\partial \alpha_2} = 0. \qquad (16.14)$$

This gives

$$J(y) \leqslant \min_{Y_2} J(Y_2) \leqslant \min_{Y_1} J(Y_1). \qquad (16.15)$$

The latter inequality in (16.15) arises because Y_2 cannot give a worse bound than Y_1 (since $Y_2 = Y_1$ for $\alpha_2 = 0$) and in general will give a better bound.

More generally, we could take a function

$$Y_n = \sum_{i=1}^{n} \alpha_i \Phi_i(x), \qquad (16.16)$$

where the Φ_i are known functions, and the parameters α_i are found from

$$\frac{\partial J}{\partial \alpha_i} = 0, \quad i = 1, \ldots, n. \qquad (16.17)$$

This will provide an upper bound for $J(y)$ and, as well, an approximation Y_n to the exact function y.

Clearly the same procedure works for a maximum principle, the approximate values of J providing lower bounds in that case.

If this process could be continued for $n \to \infty$, we should expect intuitively that under certain circumstances

$$J(y) = \lim_{n \to \infty} J(Y_n) \qquad (16.18)$$

and that

$$\lim_{n \to \infty} Y_n = y \qquad (16.19)$$

is the exact solution of the variational problem. Such a result can be proved and is the basis of all direct methods. We just state the result as follows. For the problem to make sense we must assume that there are functions in Ω for which $J(Y) < +\infty$ and also that

$$\inf_{Y} J(Y) = k > -\infty.$$

Then, there is an infinite sequence of functions $\{Y_n\}$ such that

$$\lim_{n \to \infty} J(Y_n) = k.$$

59

If the sequence $\{Y_n\}$ has a limit function y and if we can write

$$J\left(\lim_{n\to\infty} Y_n\right) = \lim_{n\to\infty} J(Y_n)$$

so that

$$J(y) = \lim_{n\to\infty} J(Y_n),$$

then

$$J(y) = k,$$

and y is the solution of the variational problem. Also, the functions of the *minimising sequence* $\{Y_n\}$ can be regarded as approximate solutions of the associated differential (Euler–Lagrange) equation problem.

EXAMPLE 16.1. Let $w = 1$, $q = 1$ in equation (16.4), and take the admissible function

$$Y_1 = \alpha x(1-x) = \alpha\Phi. \tag{16.20}$$

Putting this in (16.4) we obtain

$$J(Y_1) = \tfrac{1}{2}A\alpha^2 - B\alpha, \tag{16.21}$$

where

$$A = \int_0^1 \{(\Phi')^2 + \Phi^2\}\,\mathrm{d}x = \tfrac{11}{30},$$

and

$$B = \int_0^1 \Phi\,\mathrm{d}x = \tfrac{1}{6}.$$

The minimum value of J in (16.21) occurs for α given by $\partial J/\partial\alpha = 0$, that is

$$\alpha = \frac{B}{A} = \frac{5}{11}. \tag{16.22}$$

Putting this back in (16.21), we have

$$\min_{\alpha} J(Y_1) = -\frac{B^2}{2A} = -\frac{5}{132} = -0.037879. \quad (16.23)$$

The result (16.23) is an upper bound for $J(y)$, and

$$Y_1 = \tfrac{5}{11} x(1-x)$$

is a simple variational approximation to the exact function y

17. Complementary variational principles

The integral in equation (16.4) has the minimum property

$$J(y) \leqslant J(Y)$$

and in example 16.1 we calculated $J(Y)$ for a simple function $Y = \alpha x(1-x)$. This provided an upper bound to $J(y)$, but of course it did *not* tell us how far above $J(y)$ the bound $J(Y)$ lies. To get an accurate estimate of $J(y)$ we require more information, for example a lower bound as well so that

$$\text{lower bound} \leqslant J(y) \leqslant \text{upper bound}. \quad (17.1)$$

Then the closer these bounds, the better will our estimate of $J(y)$ be. So, the question is, can we find a lower bound in this situation?

Early in the twentieth century isolated results on this question were obtained, mainly in the theory of elasticity and electricity. Here we shall describe a unified approach to the question, one based on the *canonical theory* of section 6.

We shall discuss the problem for the integral in equation (16.4), namely

$$J(Y) = \int_0^1 \{\tfrac{1}{2}(Y')^2 + \tfrac{1}{2}wY^2 - qY\}\,dx, \quad Y(0) = Y(1) = 0.$$

$$(17.2)$$

Here

$$F = \tfrac{1}{2}(Y')^2 + \tfrac{1}{2}wY^2 - qY \qquad (17.3)$$

and we get the canonical form by introducing the variable

$$P = \frac{\partial F}{\partial Y'} = Y' \qquad (17.4)$$

and the Hamiltonian

$$H(P, Y) = PY' - F = \tfrac{1}{2}P^2 - \tfrac{1}{2}wY^2 + qY. \qquad (17.5)$$

Then, instead of $J(Y)$ in (17.2), we consider the functional

$$I(P, Y) = \int_0^1 \{PY' - H(P, Y)\}\, dx - [PY]_0^1, \qquad (17.6)$$

$$= \int_0^1 \{-P'Y - H(P, Y)\}\, dx, \qquad (17.7)$$

where the boundary term in (17.6) is added to lead to the conditions $y(0) = y(1) = 0$ (see Exercise 3 in section 20). From the canonical theory of section 6 we can see that $I(P, Y)$ is stationary at p, y where these are solutions of

$$\frac{dY}{dx} = \frac{\partial H}{\partial P} = P, \qquad (17.8)$$

$$-\frac{dP}{dx} = \frac{\partial H}{\partial Y} = -wY + q, \qquad (17.9)$$

with

$$Y(0) = 0, \quad Y(1) = 0. \qquad (17.10)$$

Now the functional $I(P, Y)$ can be used in two ways.

(i) Define

$$J(Y) = I(P(Y), Y) \qquad (17.11)$$

via (17.6), where $P(Y)$ is obtained by solving the first canonical equation (17.8), that is

$$P(Y) = Y'. \tag{17.12}$$

This gives

$$J(Y) = \int_0^1 \{\tfrac{1}{2}(Y')^2 + \tfrac{1}{2}wY^2 - qY\}\,\mathrm{d}x \tag{17.13}$$

if we impose

$$Y(0) = Y(1) = 0. \tag{17.14}$$

The functional $J(Y)$ in (17.13) is just that in (17.2) as would be expected since the procedure here takes us out of the canonical form into the Euler–Lagrange form. If

$$w \geqslant 0 \tag{17.15}$$

we know (as in sections 12 and 16) that $J(Y)$ satisfies the minimum principle

$$J(y) \leqslant J(Y). \tag{17.16}$$

Also from the definition of J in (17.11), we see that

$$J(y) = I(p, y), \tag{17.17}$$

where p and y are the critical functions, that is, solutions of (17.8)–(17.10).

(ii) The second way to use $I(P, Y)$ is to eliminate Y, by defining

$$G(P) = I(P, Y(P)) \tag{17.18}$$

via (17.7), where $Y(P)$ is the solution of the second canonical equation (17.9), that is

$$Y(P) = \frac{1}{w}(q + P'), \tag{17.19}$$

supposing $w \neq 0$. This gives

$$G(P) = -\tfrac{1}{2}\int_0^1 \left\{P^2 + \frac{1}{w}(q + P')^2\right\}\mathrm{d}x. \tag{17.20}$$

63

If p denotes the critical function satisfying (17.8) and (17.9), then (17.20) gives

$$\Delta G = G(p+\varepsilon\eta) - G(p) = -\tfrac{1}{2} \int_0^1 \left\{ (\varepsilon\eta)^2 + \frac{1}{w}(\varepsilon\eta')^2 \right\} dx,$$

the linear terms in ε on the right vanishing since $\delta G = 0$. So, if

$$w > 0, \qquad (17.21)$$

we see that the *maximum* principle

$$G(P) \leqslant G(p) \qquad (17.22)$$

holds, where p is the critical function. Also, from the definition of G in (17.18) we have

$$G(p) = I(p, y). \qquad (17.23)$$

Thus (17.22) provides a lower bound for $I(p, y)$.

Combining (17.16) and (17.22) we therefore have

$$G(P) \leqslant G(p) = I(p, y) = J(y) \leqslant J(Y) \qquad (17.24)$$

when $w > 0$. Because J and G provide opposite bounds for $I(p, y)$ they are said to give rise to *complementary variational principles*.

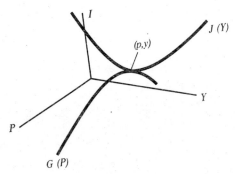

FIGURE 14

In (17.24), Y is any admissible function which satisfies $Y(0) = Y(1) = 0$, and P is any admissible function which need not satisfy any boundary condition.

Since the exact functions p and y are related by

$$p = \frac{dy}{dx},$$

see equation (17.8), it is useful to think of P in the form

$$P = \frac{dV}{dx}, \tag{17.25}$$

where V is an approximation to the function y. Then (17.24) gives

$$G(V') \leqslant I(p, y) \leqslant J(Y) \quad (w > 0). \tag{17.26}$$

Here $J(Y)$ is given by (17.13) and (17.14), $G(V')$ is determined by (17.20), and $I(p, y)$ is given by

$$I(p, y) = J(y) = -\tfrac{1}{2} \int_0^1 qy \, dx, \tag{17.27}$$

where, in deriving (17.27), we have used the fact that y satisfies (17.8)–(17.10) which are equivalent to

$$-\frac{d^2y}{dx^2} + wy = q, \quad 0 < x < 1, \tag{17.28}$$

with $y(0) = y(1) = 0$.

EXAMPLE 17.1. Let $w = 1$, $q = 1$ in (17.20), and take the admissible function

$$V_1 = \beta x(1-x) = \beta \Phi. \tag{17.29}$$

Putting this in (17.20) we obtain

$$G(V_1') = -\tfrac{1}{2} A\beta^2 - B\beta - C, \tag{17.30}$$

where

$$A = \int_0^1 \{(C\Phi')^2 + (\Phi'')^2\} \, dx = \frac{13}{3}$$

$$B = \int_0^1 \Phi'' \, dx = -2,$$

and

$$C = \tfrac{1}{2}.$$

The maximum value of G in (17.30) occurs for β given by $\partial G/\partial \beta = 0$, that is

$$\beta = -\frac{B}{A} = \frac{6}{13}. \tag{17.31}$$

Putting this back in (17.30) we have

$$\max_{\beta} G(V_1') = \frac{B^2}{2A} - C = -\frac{1}{26} = -0{\cdot}038461. \tag{17.32}$$

The result (17.32) is a lower bound for $I(p, y)$ or $J(y)$ given by (17.27) with $q = 1$. Combining this with the related upper bound result of example 16.1 in (16.23), we therefore have

$$-0{\cdot}038461 \leqslant J(y) \leqslant -0{\cdot}037879, \tag{17.33}$$

which locates the solution $J(y)$ of the variational problem to within 2 per cent. Higher accuracy can be achieved by using more elaborate variational functions such as those of the form in equation (16.13).

By solving (17.28) and using the result in (17.27) we may find the exact value of $J(y)$ in this example (see Answers to Exercises, Chapter 4, 1). However, exact solutions of variational problems are generally not available and upper and lower bounds for them are therefore of considerable value.

18. Boundary value problems

Many problems are first stated as boundary problems in differential equations, such as

$$y'' = e^y, \quad 0 < x < 1,$$

$$y(0) = 0, \quad y(1) = 1,$$

mentioned in section 16. Most of these boundary value problems cannot be solved exactly, and so it is important to develop new methods of studying them.

One method is to reformulate the boundary value problem as a variational problem, if this can be done. That is, the given differential equation is regarded as the Euler–Lagrange equation of some variational problem which is then solved by using direct methods. The variational function Y is then an approximation to the exact y, the solution of the boundary value problem. In addition, the exact functional $J(y)$ of the variational problem may be of interest and bounds for it will be useful.

Further results will be obtained if complementary principles can be found which lead to upper and lower bounds for $J(y)$.

The procedure just outlined may be summarised as follows.

Differential equation problem $D(y) = 0$. $\longrightarrow$ Identify $D(y) = 0$ as Euler–Lagrange or canonical equations.

Use direct method to obtain approximate solution Y. $\longleftarrow$ Choose corresponding functional $J(Y)$ or $I(P, Y)$.

To illustrate these ideas we consider the boundary value problem in equation (12.3)

$$-\frac{d}{dx}\left\{v\frac{dy}{dx}\right\}+wy = r, \quad 0 < x < 1, \quad (18.1)$$

with

$$y(0) = 0, \quad y(1) = 0. \quad (18.2)$$

Equation (18.1) is the Euler–Lagrange equation for the functional $J(Y)$ in equation (12.1). However, let us try to get a canonical formulation which will then give complementary functionals J and G.

We rewrite (18.1) as the pair

$$\frac{dY}{dx} = \frac{1}{v}P = \frac{\partial H}{\partial P}, \quad (18.3)$$

$$-\frac{dP}{dx} = -wY+r = \frac{\partial H}{\partial Y}, \quad (18.4)$$

with solution p, y. A suitable Hamiltonian H is

$$H(P, Y) = \frac{1}{2v}P^2 - \tfrac{1}{2}wY^2+rY. \quad (18.5)$$

We have thus identified the differential equation (18.1) as canonical Euler equations (18.3) and (18.4), and the next step is to write down the corresponding functional $I(P, Y)$. This was given in section 17 for the boundary conditions (18.2), and is

$$I(P, Y) = \int_0^1 \{PY' - H(P, Y)\}\,dx - \left[PY\right]_0^1 \quad (18.6)$$

$$= \int_0^1 \{-P'Y - H(P, Y)\}\,dx. \quad (18.7)$$

68

Using the Hamiltonian H in (18.5) and following the procedure of section 17, we find that the functionals J and G are

$$J(Y) = \int_0^1 \left\{ \tfrac{1}{2}(Y')^2 v + \tfrac{1}{2} wY^2 - rY \right\} \mathrm{d}x, \quad Y(0) = Y(1) = 0, \quad (18.8)$$

and

$$G(P) = -\tfrac{1}{2} \int_0^1 \left\{ \frac{1}{v} P^2 + \frac{1}{w}(r+P')^2 \right\} \mathrm{d}x \quad (v \neq 0). \quad (18.9)$$

Taking the case $v > 0$ and $w > 0$, we see that the complementary principles

$$G(P) \leqslant I(p, y) \leqslant J(Y) \quad (18.10)$$

hold.

Since the exact p and y are related through $p = v\,\mathrm{d}y/\mathrm{d}x$, by equation (18.3), it is useful to take

$$P = v\frac{\mathrm{d}V}{\mathrm{d}x}, \quad (18.11)$$

where V is an approximation to y.

The final step is to minimise $J(Y)$ and maximise $G(P)$ for suitable choices of functions, as described in section 16. This will provide upper and lower bounds for $I(p, y)$, which for (18.1) and (18.2) is given by

$$I(p, y) = -\tfrac{1}{2} \int_0^1 ry \, \mathrm{d}x. \quad (18.12)$$

In addition, Y will be an approximation to the exact function y.

This approach to boundary value problems dates from the 1960s and has very wide application.

19. Rayleigh bound for eigenvalues

Eigenvalue problems are concerned with the search for numbers λ which arise in equations like

$$Ly = \lambda y, \tag{19.1}$$

where L is some operator and y is subject to some boundary conditions. For example, we might be looking for the eigenvalues of

$$-\frac{d^2y}{dx^2} = \lambda y, \quad 0 < x < 1, \tag{19.2}$$

with

$$y(0) = y(1) = 0. \tag{19.3}$$

Now this eigenvalue problem is simple enough to be solved exactly since

$$y = A \sin \sqrt{(\lambda)}x + B \cos \sqrt{(\lambda)}x \tag{19.4}$$

is the general solution of (19.2), and the boundary conditions (19.3) imply

$$B = 0, \quad \text{and} \quad \sin \sqrt{(\lambda)} = 0, \quad \text{i.e.} \quad \sqrt{(\lambda)} = n\pi, \quad n = 1, 2, \ldots \tag{19.5}$$

So

$$\lambda = n^2\pi^2, \quad n = 1, 2, \ldots. \tag{19.6}$$

Equation (19.6) gives the possible eigenvalues, and (19.4) gives the corresponding eigenfunctions $y_n = A_n \sin n\pi x, n = 1, 2, \ldots$.

In general, eigenvalue problems cannot be solved exactly, and it is therefore desirable to have some approximate method for estimating eigenvalues. We shall illustrate one such method, due to Rayleigh, by taking the problem in (19.2) and (19.3). Let λ_1 be the lowest eigenvalue of this problem, and suppose Ω is the class of functions with continuous second derivatives in $(0, 1)$ which vanish at the end points. Then it follows that

$$J(Y) = \int_0^1 Y \left(-\frac{\mathrm{d}^2}{\mathrm{d}x^2} - \lambda_1 \right) Y \, \mathrm{d}x \geqslant 0 \quad \text{for all} \quad Y \in \Omega.$$

$$(19.7)$$

To see this, expand $Y \in \Omega$ in a series of eigenfunctions $y_n = A_n \sin n\pi x$ of $-\mathrm{d}^2/\mathrm{d}x^2$ which are zero at $x = 0$ and $x = 1$,

$$Y = \sum_{n=1}^{\infty} a_n y_n.$$

Then

$$J(Y) = \int_0^1 \sum_{m=1}^{\infty} a_m y_m \sum_{n=1}^{\infty} (\lambda_n - \lambda_1) \, a_n y_n \, \mathrm{d}x.$$

Since

$$\int_0^1 y_n y_m \mathrm{d}x = 0 \quad \text{for} \quad n \neq m$$

we have

$$J(Y) = \sum_{n=1}^{\infty} a_n^2 (\lambda_n - \lambda_1) \int_0^1 y_n^2 \, \mathrm{d}x$$

$$\geqslant 0,$$

since λ_1 is the *lowest* eigenvalue.

Rewriting (19.7) we have

$$\lambda_1 \leqslant \frac{\left\langle Y, -\dfrac{\mathrm{d}^2 Y}{\mathrm{d}x^2} \right\rangle}{\langle Y, Y \rangle} = \Lambda(Y) \quad \text{say,} \qquad (19.8)$$

where $\langle Y_1, Y_2 \rangle = \int_0^1 Y_1 Y_2 \, \mathrm{d}x$. Thus $\Lambda(Y)$ provides an upper bound for the lowest eigenvalue λ_1. This is the Rayleigh upper bound. Clearly $\Lambda(y_1) = \lambda_1$, where y_1 is the eigenfunction of the lowest eigenvalue.

EXAMPLE 19.1. As a simple example of the Rayleigh bound we take $Y = x(1-x)$ in (19.8) and find that

$$\Lambda(Y) = \frac{1/3}{1/30} = 10,$$

which is only slightly greater than the exact value $\lambda_1 = \pi^2 \simeq 9\cdot87$ as given by (19.6).

More generally, if the eigenvalue problem is given by (19.1) with suitable boundary conditions, the Rayleigh bound for λ_1 is

$$\lambda_1 \leqslant \frac{\langle Y, LY \rangle}{\langle Y, Y \rangle}. \tag{19.9}$$

Lower bounds for eigenvalues can also be obtained in some cases, but the theory is somewhat beyond the scope of this book.

20. Exercises

1. Calculate the exact value of $J(y)$ in example 16.1.

2. Use the Rayleigh–Ritz method to find an approximate solution of the problem of minimising the functional

$$J(y) = \tfrac{1}{2} \int_0^2 (y'^2 + y^2 + 2xy)\,\mathrm{d}x, \quad y(0) = y(2) = 0,$$

and compare the result with the exact solution.

3. Show that the functional $I(P, Y)$ in equations (17.6) and (17.7) is stationary at (p, y), where p and y satisfy equations (17.8)–(17.10).

4. Show that the problem complementary to that in Exercise 2 involves maximising the functional

$$G(\psi') = -\tfrac{1}{2} \int_0^2 \{(\psi')^2 + (\psi'' - x)^2\}\, dx,$$

and use this to generate a second approximate solution.

5. Derive complementary variational principles for the differential equation boundary value problem

$$\frac{d^2y}{dx^2} - k^2 y = 1, \quad 0 < x < a,$$

$$y(0) = y(a) = 0.$$

Show that upper and lower bounds are obtained for $\int_0^a y\, dx$ where y is the solution of the boundary problem.

6. Let $L = d^4/dx^4$ be defined for the class of functions y with continuous derivatives of fourth order on $0 < x < 1$, which satisfy $y(0) = y(1) = y'(0) = y'(1) = 0$. Use the Rayleigh method to obtain an upper bound for the lowest eigenvalue of L.

Answers to Exercises

Chapter One

1. (i), (ii) Any admissible y.
   (iii) No continuous critical curve.

2. (i) $y = \frac{1}{2} \sin x + Ae^x + Be^{-x}$.
   (ii) $y = -\frac{1}{2}x \cos x + A \cos x + B \sin x$.
   (iii) $y = \frac{1}{2} \cosh x + A \cos x + B \sin x$.
   (iv) $y = \frac{1}{2}xe^x + Ae^x + Be^{-x}$.

3. $y = A \int \{f^2 - A^2\}^{-1/2} \, dx + B$.

   (i) $f = x^{1/2}$, $y = 2A(x - A^2)^{1/2} + B$.
   (ii) $f = x$, $y = A \cosh^{-1}(x/A) + B$.

4. (i) $y = A \left(1 + \dfrac{x^2}{4A^2}\right)$, $A = \frac{1}{2}b \pm \frac{1}{2}(b^2 - 1)^{1/2}$.

   One solution for $b = 1$, two for $b > 1$, and none for $b < 1$.
   (ii) $y = \sinh(Cx + D)$.

Chapter Two

1. $y = z = \sinh x / \sinh \pi/2$.

3. $\left(\dfrac{\partial S}{\partial x}\right)^2 + \left(\dfrac{\partial S}{\partial y}\right)^2 = f^2(y)$.

74

Critical curves $x - \alpha \int^y \{f^2(\lambda) - \alpha^2\}^{-1/2} \, d\lambda = \beta$,

where α and β are constants.

4. $\int (1 + y'^2)^{1/2} \, dx$.

5. Lagrange equation $m\ddot{x} = -kx$.

   Hamilton equations $\dfrac{dx}{dt} = \dfrac{p}{m}$, $\quad -\dfrac{dp}{dt} = kx$.

6. Lagrange equations $mr\dot{\theta}^2 - \dfrac{k}{r^2} - m\ddot{r} = 0$,

$$\frac{d}{dt}(mr^2\dot{\theta}) = 0.$$

$H = \dfrac{1}{2m}\left(p_r^2 + \dfrac{1}{r^2}p_\theta^2\right) - \dfrac{k}{r}$, where $p_r = \dfrac{\partial L}{\partial \dot{r}}$, $p_\theta = \dfrac{\partial L}{\partial \dot{\theta}}$.

Hamilton equations $\dfrac{dr}{dt} = \dfrac{p_r}{m}$, $\quad -\dfrac{dp_r}{dt} = -\dfrac{p_\theta^2}{mr^3} + \dfrac{k}{r^2}$,

$$\frac{d\theta}{dt} = \frac{p_\theta}{mr^2}, \quad -\frac{dp}{dt} = 0.$$

Chapter Three

1. $y = x^3$.

2. $y = 7 - \dfrac{4}{x}$.

4. $y = \tfrac{1}{3}e^{2x}$.

5. $y = \sum\limits_{n=1}^{\infty} a_n \sin n\pi x$, where $\sum\limits_{n=1}^{\infty} a_n^2 = 4$.

6. $y = x^2$.

7. $y = \left(\dfrac{1}{2\pi\sigma^2}\right)^{1/2} \exp\left\{-\dfrac{x^2}{2\sigma^2}\right\}$, i.e. the normal distribution.*

8. By Euler's rule $y = x$. But for this, $K(y) = \sqrt{2} \neq 2$. That is, the integral constraint is *not* satisfied. The reason for this situation is that $y = x$ is a critical curve of $K(y)$.

Chapter Four

1. $J(y) = -\dfrac{1}{2} + \dfrac{e-1}{e+1} = -0.037883$.

2. For $Y = \alpha x(2-x)$, $\min\limits_{\alpha} J(Y) = -\dfrac{5}{21} = -0.238$, at $\alpha = -\dfrac{5}{14}$.

   Exact $J(y) = -0.259$.

4. For $\psi = \beta x(2-x)$, $\max\limits_{\beta} G(\psi') = -\dfrac{7}{12} = -0.583$, at $\beta = -\dfrac{3}{8}$.

6. For $Y = x^2(1-x)^2$, upper bound is 504.
   Exact value is slightly larger than $(3\pi/2)^4 \approx 500$.

---

* See, for example, A. M. Arthurs, *Probability Theory*, Routledge & Kegan Paul (Library of Mathematics), London (1965).

Suggestions for Further Reading

ARTHURS, A. M., *Complementary Variational Principles*, Clarendon Press, Oxford, 1970.

BELLMAN, R., *Dynamic Programming*, Princeton University Press, 1957.

GELFAND, I. M. and FOMIN, S. V., *Calculus of Variations*, Prentice-Hall, Englewood Cliffs, New Jersey, 1963.

GOULD, S. H., *Variational Methods for Eigenvalue Problems*, 2nd edition, University of Toronto Press, 1966.

LANCZOS, C., *The Variational Principles of Mechanics*, 3rd edition, University of Toronto Press, 1966.

MOISEIWITSCH, B. L., *Variational Principles*, Interscience Publishers, New York, 1966.

YOUNG, L. C., *Lectures on the Calculus of Variations and Optimal Control Theory*, Saunders, Philadelphia, 1969.

Index